W9-AUS-830

HISTORICAL FIRST PATENTS
The First United States Patent for Many Everyday Things

by
TRAVIS BROWN

The Scarecrow Press, Inc.
Metuchen, N.J., & London
1994

British Library Cataloguing-in-Publication data available

Library of Congress Cataloging-in-Publication Data

Brown, Travis, 1926-
Historical first patents : the first United States patent for many everyday things / by Travis Brown
p. cm.
Includes bibliographical references and index.
ISBN 0-8108-2898-7 (alk. paper)
1. Patents—United States—History. 2. Inventions—United States—History.
I. Title.
T223.P2B76 1994
608.773—dc20 94-14814

Manufactured in the United States of America
Printed on acid-free paper

To my wife Maxine and
my son David,

both of whom God gave me
for a more abundant life.

Acknowledgements

I am grateful to Joseph Kane's book *Famous First Facts* and to the biographical series *Dictionary of American Biography*. These books originally gave me the idea and the motivation to write this book.

I am also grateful to the following individuals and organizations: Linda Blanton, Gail Campbell, John Hart, Al Maupin, and library director Mark Thomas of the Johnson City Public Library, Johnson City, Tennessee, whose help was invaluable in securing many of the references; Charles Van Horn, Kenneth Downey, Mary Downey, Norman Torchin, and Charles Bowers Jr. of the United States Patent and Trademark Office, whose encouragement and help gave me the perseverance necessary to finish this project; Oscar Mastin, public affairs officer of the United States Patent and Trademark Office, for some valuable advice concerning technical matters; Wilma Reister of the Area Resource Center of Tennessee, Knoxville, Tennessee, for providing some hard-to-find biographies; Claudia Gaskins for typing some early manuscript pages; the Scientific Library of the United States Patent and Trademark Office, where many of the early references used to collect the facts in this book were found; the National Archives that provided some patents granted before 1836; Julie and Allen Groseclose, Black Cats Press, Jonesborough, Tennessee, "Pete" Peterson, MCS, Inc., Johnson City, Tennessee, Steve McKinney, Computer Works, Johnson City, Tennessee, and Keith Rodgers, for some valuable computer assistance.

I also want to thank Jane Hillhouse of Hillhouse Graphic Design, Kingsport, Tennessee, for the layout work of the book and Jill Oxendine and Shirley Covington for proofreading the manuscript.

Finally, special thanks go to my wife, Maxine, who put up with a cluttered work space while this work was in progress.

Travis Brown, May, 1994
Johnson City, Tennessee

Table of Contents

Introduction

This is a book of "firsts"—the first United States patent granted for many common, everyday things. Although any book of "firsts" naturally brings up the argument of who really is the first, it is not the author's intent that this work be definitive, but that it may at least serve as a background for further research. The patents selected as "firsts" are the result of the author's own research and do not officially reflect the opinion of any agency or organization, especially that of the United States Patent and Trademark Office.

One of the purposes of the book is to acquaint the reader with a small section of the American history of technology as exemplified by the granting of United States patents for inventions, some of which have notably influenced the economy and civilization of the United States and the world. An attempt has been made to fulfill this purpose by including for each invention a short history of the invention, a brief biographical sketch of the inventor, and the steps the inventor took in making the invention. In some cases, the legal and administrative procedures have been included as well as the impact the invention had, and continues to have, on the marketplace. Of course, space would not permit including every invention, but those included represent a fair cross-section of those that continue to influence the technological advances of the United States, and, in most cases, the world. It also must be kept in mind that because this book is about first patents issued for many inventions, modern-day devices offer much improved versions, these improvements being the subject of patents themselves.

A bibliography has been included at the end of the book for those who want further information or background on a particular invention. Many of the references date back to the time of the invention or shortly thereafter. A copy of the patent drawing is included with the text for each invention. For those patents that contain no drawing, the first page of the specification is included.

A patent is a contract between the United States government and an inventor that gives to the inventor certain property rights for 17 years in exchange for disclosing to the public how to make the invention. For this to be accomplished, there must be a set of laws governing this action and an agency of the government that performs this duty. Congress has set forth certain patent laws and has created an office for administering these laws. Thus, brief histories of the United States patent laws and the United States Patent and Trademark Office are included. From 1790 to 1880, the United States Patent Office required a miniature model to be submitted with every patent application. Because these models represent an important part of our American heritage, a section on these models has also been included.

United States Patent Laws

The first patent statute was instituted in the Republic of Venice in 1474. The statute awarded a ten-year monopoly to anyone who introduced new technology into Venice. Monopolies were granted in England during the reign of Queen Elizabeth I and continued during the reign of King James I. Many of these monopolies were given as rewards to court favorites and not as a result of any new technology. In 1623, during the reign of James I, Parliament passed the Statute of Monopolies, because there was so much protest against this practice of rewarding court favorites. This statute abolished all monopolies except inventions, and limited the term to 14 years.

The American colonies followed the practice set forth in England and granted exclusive rights to both inventors and those introducing new industry in the

New World. The first American patent was granted to Samuel Winslow in 1641 by the Massachusetts Bay Colony for a novel method of making salt. Other colonies that followed Massachusetts were Virginia in 1652, South Carolina in 1691, New York in 1712, Connecticut in 1716, Rhode Island in 1731, and Maryland in 1770. Some colonies did not grant patents until after the Revolutionary War. These were New Hampshire in 1786, Georgia in 1788, Delaware in 1788, New Jersey in 1788, and Pennsylvania in 1789. Of the 13 original colonies, only North Carolina did not grant any patents. The last state to grant a patent was New Hampshire in 1791.

The colonists, angered by the British Parliament's passage of the Stamp Act, the Townsend Revenue Act, the Tea Act, and the Intolerable Acts, began to think and act in terms of independence. On September 5, 1774, the First Continental Congress met in Philadelphia for two months and issued a declaration of ten "rights." On June 11, 1776, Thomas Jefferson was asked to produce a Declaration of Independence. On July 4, 1776, this Declaration of Independence was approved and signed by Congress President John Hancock.

On May 14, 1787, delegates from the various states met in Philadelphia to frame the Constitution of the United States. On August 18, 1787, James Madison of Virginia and Charles Pinckney of South Carolina submitted proposals for the protection of authors and inventors with patents. On September 17, 1787, this Constitution was signed by the delegates, including Article I, Section 8, which states that Congress has the power "to promote the progress of science and useful arts by securing for limited times to authors and inventors the exclusive right to their respective writings and inventions." On March 4, 1789, the government under the new Constitution began.

Noah Webster, of dictionary fame, was one of the first to write a bill for the American patent system. On April 16, 1789, he wrote what he called a federal copyright bill, a combined patent and copyright bill. A committee was appointed to draft a general law. On June 23, 1789, House Bill 10, a printed document of 11 pages, was presented as a combined federal copyright and patent bill. After much argument, this bill was defeated. Congress then, on January 15, 1790, appointed Edmund Burke of South Carolina, Benjamin Huntington of Connecticut, and Lambert Cadwallader of New Jersey, as a committee for the purpose of bringing in a federal patent and copyright bill. A short time later, the committee brought forth House Bill 41. The House, after amending to provide for no examination but publication in newspapers and opposition before the Secretary of State, passed the bill on March 10, 1790, and sent it to the Senate. The Senate made 12 changes in the bill on March 30, 1790. The major change was the deletion of the publication and opposition procedures, and the insertion of the requirement of an examination before a Patent Board, consisting of the Secretary of State, the Secretary of War, and the Attorney General. The President, upon the granting of an application by the Patent Board, affixed the seal of the United States to the document. Another important change was that the patent law was separated from the copyright law. This amended bill passed and was signed by President Washington on April 10, 1790, and became the first United States patent law. On May 31, 1790, Congress enacted the first United States copyright law.

The Patent Board, consisting of Secretary of State Thomas Jefferson, Secretary of War Henry Knox, and Attorney General Edmund Randolph, could not devote much time to patent matters due to other important work. This prompted Senator Williamson of South Carolina, with the help of Thomas Jefferson, to introduce a new patent bill in Congress on February 7, 1793. Congress enacted the Patent Act of 1793 on February 21, 1793. This bill eliminated an examination by the Patent Board and substituted a system of registration administered by the Department of State.

The influx of petitions for patents increased to such an extent that Secretary of State James Madison created by decree in 1802 a separate Patent Bureau within the Department of State. Dr. William Thornton was appointed as the first superintendent. Dr. Thornton was born in 1759 at Tortola, British West Indies, and received a medical degree at Edinburgh, Scotland. He came to the United States in 1787 and became a citizen in 1788. Although untrained as an architect, he designed the first Capitol building in Washington, for which he received $500 and a plot of land. Dr. Thornton served as superintendent of patents from 1802 until his death in 1828.

For more than 40 years, the Department of State issued patents on every application without any examination of the merits or novelty of the invention. The need for reform of the patent laws prompted John Ruggles, a newly elected Senator from Maine, on December 31, 1835, to propose that the Senate appoint a committee to recommend revisions in the patent laws. The Senate promptly appointed him as chairman of the committee. On April 28, 1836, the committee submitted its report, together with Senate Bill 239. Congress passed the Patent Act of 1836 on July 4, 1836. This new law restored the examination of applications, created a Patent Office, established a Scientific Library, and appointed Henry L. Ellsworth as the first Commissioner of Patents. Another law passed on the same day provided for the erection of a new Patent Office building.

Designs were made patentable for a term of seven years on August 29, 1842. The Patent Office was transferred from the State Department to the newly created Department of Interior on March 3, 1849. On February 5, 1859, copyrights were transferred to the Department of Interior with the Commissioner of Patents placed in charge. Congress passed the Patent Act of 1861 on March 2, 1861, which increased the term of a patent from 14 years to 17 years. This act authorized the printing of 10 copies of the description and claims and 10 copies of the drawing. The practice was discontinued after nine months because of the Civil War but was started again on November 29, 1866. The Patent Act of 1861 also established the Board of Appeals.

On May 21, 1861, the Confederate States of America established their own Patent Office. Rufus R. Rhodes of Louisiana, former examiner in the Washington Patent Office, became the first and only Patent Commissioner. The Confederate Patent Office was located in the Mechanics Building in Richmond, Virginia.

Congress consolidated the 25 patent acts passed during the past 34 years by enacting the Patent Act of 1870 on July 8, 1870. This act contained several provisions, the most important being the first law concerning trademarks, models were no longer required except as requested by the commissioner, and copyrights were transferred to the Library of Congress. However, on June 18, 1874, Congress authorized the Patent Office to register all copyrights for prints and labels for articles of manufacture. This registration continued until July 1, 1940 when the registration of copyrights was transferred from the Patent Office to the Library of Congress, thus ending the connection of the Patent office with copyright matters.

On June 10, 1898, Congress authorized the establishment of a classification division in the Patent Office. The Trademark Act was passed February 20, 1905, which authorized the Patent Office to register trademarks used in interstate commerce. On April 1, 1925, the Patent Office was transferred from the Department of Interior to the Department of Commerce where it still remains. The Patent Act of May 23, 1930, authorized plants to be patented. On July 19, 1952, President Harry Truman signed a new patent act which revised and codified all patent laws that became effective on January 1, 1953, as Title 35 of the United States Code.

Two other important legislative acts since 1953 are the Patent Cooperation Treaty and the re-examination procedure. The Patent Cooperation Treaty, ratified by Congress on November 26, 1975, is an international agreement that provides for the filing of patent applications on the same invention in several party countries. On July 1, 1981, Public Law 96-517 went into effect. This law allows any person to call to the attention of the Patent and Trademark Office any prior art that may have a bearing on the patentability of any claim in any unexpired patent that was issued before as well as after that date.

United States Patent Office

After Congress passed the first Patent Act on April 10, 1790, the first home of the Patent Office, though the bill created no Patent Office as such, was in New York City. It was here that the Secretary of State, the Secretary of War, and the Attorney General issued, and the President signed, the first United States patent to Samuel Hopkins. The patent was for a new process for making potash, an ingredient useful in soap and fertilizer manufacture. Only one other patent was issued to Joseph Simpson during the five months the government remained in New York.

Late in 1790, the seat of government was moved to Philadelphia and remained there until 1800. While in Philadelphia, the Patent Board, consisting of Secretary of State Thomas Jefferson, Secretary of War Henry Knox, and Attorney General Edmund Randolph met as time allowed to examine and pass to issue those applications they deemed to merit invention. In 1790, they issued three patents, two in New York and one to Oliver Evans in Philadelphia. They issued 33 in 1791 and 11 in 1792. In 1793, prior to February 21, they issued 10. On February 21, 1793, Congress passed the Patent Act of 1793 where the system of obtaining patents went from an extreme rigid examination by high government officials to a system of registration by a clerk in the State Department. During the Patent Act of 1790, a total of 57 patents were issued. Under the Patent Act of 1793, while the government remained in Philadelphia, 235 patents were issued, including Eli Whitney's cotton gin.

In 1800, the seat of government was moved from Philadelphia to Washington, D.C. The State Department that handled patent matters was assigned temporary quarters for a month in the Treasury Office before moving into the "Seven Buildings" between 19th and 20th Streets on Pennsylvania Avenue. In May of 1801, the State Department with the attached Patent Office moved into the new Southwest Executive Building, also known as the War Office, sharing the space with the War Department, the Navy, the Post Office, and the city government.

Due to an increase in the number of patent applications filed, the Jefferson administration in May of 1802 organized the Patent Bureau and appointed Dr. William Thornton as head. A short time later, the Bureau moved into "Croker's two-story house" on 8th Street between E and F Streets.

On April 28, 1810, Congress authorized $10,000 for the purchase of a building for the Patent Bureau and the General Post Office. Blodgett's Hotel on the north side of E Street between 7th and 8th Streets was selected and purchased for this purpose. The building was built in 1793 by Samuel Blodgett but had never been finished. A portion of it housed the United States Theatre, the first formal playhouse in Washington. Congress authorized an additional $3,000 for finishing and repairing the building. The Patent Bureau and the General Post Office moved into this building in 1811. Due to the efforts of Dr. Thornton, it was the only government building that was not burned by the British in 1814. Congress met in this building in 1815 while the Capitol was being restored. Due to inadequate space for the patent models and records, an addition was built facing 7th Street in 1829.

On July 4, 1836, Congress authorized funds for the construction of a new building for the Patent Office. However, before construction had proceeded very far, a fire on December 15, 1836 destroyed the building housing the Patent Office and General Post Office. Over 10,000 records of patented inventions and 7,000 patent models were destroyed. After the fire, the Patent Office was given some rooms in City Hall, located at the head of John Marshall Place and facing south towards Pennsylvania Avenue, where it remained until the F Street wing of the new building was finished in 1840.

On March 3, 1837, Congress appropriated $100,000 for the purpose of duplicating the records destroyed in the fire. During the first year after the fire, about 2,000 patents were restored. This number decreased to such an extent that about 10 years later, systematic efforts to secure records ceased.

In 1840, the Patent Office moved into their new building on F Street. The need for additional space forced the construction of the east wing on 7th Street in 1852, the west wing on 9th Street in 1856, and the north wing on G Street in 1867. The total cost of the completed building was about $3 million. During the Civil War, the model rooms were used as a hospital that sheltered some 800 wounded soldiers at the end of 1862. On September 24, 1877, another fire broke out on the third floor of the G and 9th Street wings, destroying many more records and models.

On March 3, 1849, the Department of Interior was created and the Patent Office was transferred from the State Department. On April 1, 1925, the Patent Office was transferred to the Commerce Department, and in 1932 moved to the Commerce Building on 14th Street. Due to World War II, the Patent Office moved to 900 Lombardy Street in Richmond, Virginia, in 1942. After the war, the Office moved back to Washington, first to temporary quarters at Gravelly Point, and then in 1947 to the Commerce Building. In 1967 and 1968, the Patent Office moved to Crystal City, Virginia, just across the Potomac River from Washington in Northern Virginia, where it is presently located.

United States Patent Models

On April 10, 1790, the first Patent Board, made up of Secretary of State Thomas Jefferson, Secretary of War Henry Knox, and Attorney General Edmund Randolph, decided that a miniature working model of each invention should be submitted as part of the application. This requirement continued until 1870 when Congress enacted a law that no longer required a model except as requested by the Commissioner of Patents. In 1880, the requirement was dropped altogether except for flying machines and perpetual motion machines. In 1903, after the Kitty Hawk experiment, the requirement for flying machines was dropped, but the requirement for perpetual motion machines remains in effect to this day.

From 1790 to 1880, more than 200,000 models were received by the Patent Office. Fires in 1836, 1877, and 1943 destroyed thousands of models. Inadequate storage facilities resulted in the destruction of many more. Today, it is estimated that less than 100,000 remain in museums and private collections.

From the beginning the models, although limited in size to a maximum of 12 inches on a side, presented a problem as far as storage was concerned. Their numbers increased to such an extent that lack of storage space was probably one of the main reasons for Congress moving the Patent Office and model collection from a crowded office in the State Department to Blodgett's Hotel in 1811. Here, the models became a major point of interest for those visiting Washington. By 1823, the Patent Office had 1,819 models and by 1836, the number had increased to 7,000.

On July 4, 1836, Congress passed a law allowing for the construction of a new building to house the Patent Office and the model collection. However, before this could be accomplished, a fire on December 15, 1836, destroyed all records and models. Congress appropriated $100,000 for the restoration of the most important models destroyed in the fire. Due to apathy on the part of the inventors who were to send in replacement models, only a little more than $25,000 was spent on restoration of these models. However, this did not deter inventors from presenting new ones for new inventions. In fact, shortly after moving into their partially completed new building, the Patent Office became inundated by these new models. In 1856, the storage and exhibiting situation was partially solved by the addition of extra space to house the models. The model collection again became a major tourist attraction.

By 1876, the models had again multiplied to such an extent that the public was barred from the exhibit due to lack of space for the estimated 175,000 models. On September 24, 1877, fire again struck the Patent Office and destroyed 76,000 models. Congress appropriated $45,000 for replacement, but few were replaced. However, the fire did not stop the inflow of new models. By 1890, there were so many models that the Patent Office did not have room to even store them much less exhibit them. Congress finally remedied the situation by passing a bill that no longer required the submission of a model except in special cases.

This law did nothing to alleviate the storage problem that still existed at the Patent Office. Finally, in 1893, Congress rented the Union Building at G Street between 6th and 7th Streets for a model exhibition hall. The models were not properly exhibited so had few visitors. The rent of $10,000 a year was a constant source of trouble. In 1907, when the owners tried to raise the rent, Congress decided to rid themselves of the problem by selling all the models. Before selling, Congress gave the Smithsonian Institution six months to select the ones they deemed to be of historical value. In the short time available, the Smithsonian selected 1,061. In 1908, about 3,000 models for inventions that had failed to receive patents were sold at public auction for $62.18. The remaining models, approximately 155,000, were stored in various warehouses in Washington, many of them in an unused tunnel near the House Office Building on Capitol Hill.

On February 13, 1925, after learning that $200,000 had been spent for storage of patent models since 1884, Congress appropriated $10,000 to sell the models to save storage costs. A commission, consisting of the Commissioner of Patents, an official of the Smithsonian Institution, and a patent attorney, was created to select appropriate models for the Smithsonian and other museums and to dispose of the remaining models. The Smithsonian selected 2,500, and inventors and other museums, including the Henry Ford Museum of Dearborn, Michigan, selected another 2,600. The remaining models were prepared for the auction block.

In 1926, Sir Henry Wellcome, founder of a pharmaceutical company, bought the collection for about $8,000 with the idea of building a patent model museum. He had the collection moved from Washington to a Burroughs, Wellcome Company warehouse in Tuckahoe, New York. The stock market crash of 1929 and Sir Henry's death in 1936 eliminated the idea of a museum. The trustees of his estate, two years after his death, sold the collection for $50,000.

The new owners were Crosby Gaige, a Broadway producer, and Douglas G. Hertz, former owner of a New York football team. They rented a floor in the International Building of Rockefeller Center and displayed many of the models, including the original model of the Gatling gun. Although the models were page one news, few were sold until October, 1938, when Gaige and Hertz sold the entire collection to a group of businessmen for $75,000.

These businessmen incorporated under the name of American Patent Models. Many of the models were shipped and stored at the Neptune Storage Warehouse in New Rochelle, New York. Some models were sent to be displayed in showrooms in Minneapolis, Salt Lake City, and Oakland, California. For the next three years, the owners tried without success to interest some museum or individual in buying the collection and completing the museum idea of Sir Henry Wellcome. In 1941, Neptune Storage filed a lien against American Patent Models for unpaid storage bills. In an effort to raise money, American Patent Models reduced the price on all models to one dollar. An unspecified number of models were sold to many unnamed buyers. However, this effort did not raise sufficient funds, so American Patent Models took bankruptcy in 1942, reporting a loss of more than $100,000. In May 1942, the remaining models were sold at public auction in the United States Custom House in New York. Another group of businessmen bought the collection for $2,100, plus storage charges of $11,614. One of the buyers was O. Rundle Gilbert, a New York auctioneer and an authority on Americana.

Gilbert and his partners immediately moved the model collection to several warehouses on his estate in Garrison, New York. On April 14, 1943, about 2,000 models went on display at the Architectural

League in New York City. Only 400 were sold, and the rest were transported back to Garrison resulting in a net loss of $3,000. Also in 1943, another 20,000 models were lost due to fire in one of the warehouses in Garrison.

In 1949, Gilbert and his partners purchased a barn in Center Sandwich, New Hampshire. They remodeled it into a museum, displayed about 1,000 of the most interesting models, and charged one dollar for admission. In 1950, Gimbels Department Stores in New York and Philadelphia invited the owners to feature for sale a display of their patent models in their stores. There were many lookers but few buyers with only about 600 models sold. This discouraged the partners so much that they sold their interest to Gilbert at cost. In 1952, he purchased an abandoned hospital at Plymouth, New Hampshire, converted it into a museum, and moved his exhibit from Center Sandwich.

In 1970, after owning the models for 30 years and spending over $85,000, Gilbert decided to sell them. He placed his son, Robert W. Gilbert, in charge of the sale. Robert held periodic sales that attracted as many as 1,500 buyers at one time. In 1979, the remaining models, now numbering about 40,000, were sold to Cliff Petersen, a California aerospace engineer, for an undisclosed amount.

In 1992, Petersen donated his model collection and $1 million to a recently established non-profit public organization called the United States Patent Model Foundation. The foundation's objective is to recover as many models as possible in private hands and to convey them to the Smithsonian Institution's National Museum of American History. Their goal is to raise $20 million so the Smithsonian can construct a building at Silver Hill, Maryland, where these models can be restored, housed, and prepared for exhibition and research.

AC Induction Motor

The electricity that runs the motors of many appliances in the modern home usually travels from distant power plants. The man most responsible for this is Nikola Tesla, who received the first U.S. patent, No. 381,968, on May 1, 1888, for an alternating current induction motor and the first U.S. patent, No. 382,280, on May 1, 1888, for the polyphase system of power transmission.

Electric motors are grouped into two classes, those that derive their power from direct current, dc motors, and those that derive their power from alternating current, ac motors.

To understand the importance of the ac motor, a history of the dc motor is included here. In 1821, Michael Faraday of England demonstrated that rotary motion could be obtained from a magnet and a movable wire. In 1822, Englishman Peter Barlow applied Faraday's principle to obtain the rotary motion of a wheel. In 1831, Joseph Henry, an American, invented the electromagnet and from it built a crude seesaw electric motor. In 1837, American Thomas Davenport received the first U.S. patent, No. 132, for a dc electric motor. In 1873, Belgium's Zenobe Gramme and his associate, Hippolyte Fontaine of France, demonstrated that the dc generator was reversible and could be used as a dc motor if driven by another dc generator. In 1884, Gaulard of France and Gibbs of England showed that with a single phase transformer, current could be transported for miles. In 1884, American John Hopkinson, demonstrated that an ac generator could be run as a motor and was probably the first synchronous motor. In 1885, another American, William Stanley, built improved Gaulard and Gibbs transformers that were used in the first alternating current station built by Westinghouse in Buffalo, New York, in 1886. In 1888, Italian Galileo Ferraris discovered that a single current could be split into two out-of-phase currents that produced two out-of-phase magnetic fields. These magnetic fields produced a single resultant constant rotating magnetic field. Also in 1888, others who worked on ac current and induction motors were Americans Oliver B. Shallenberger and Elihu Thomson, Swiss-born Charles E. L. Brown and Russian-born Michael von Dolivo-Dobrowolsky, both of Germany. But it was Nikola Tesla who is credited

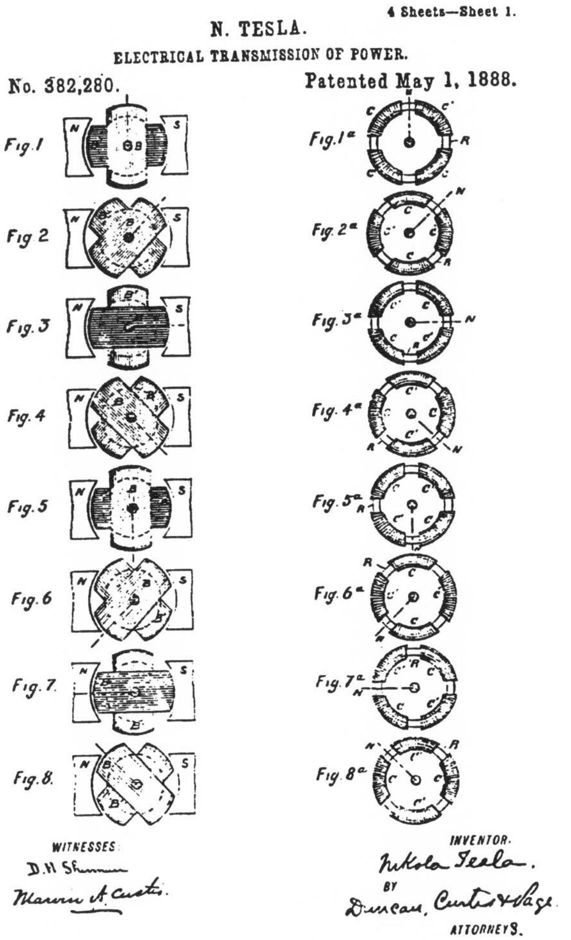

4 Sheets—Sheet 1.
N. TESLA.
ELECTRICAL TRANSMISSION OF POWER.
No. 382,280. Patented May 1, 1888.

Nikola Tesla, received the first U.S. patent, No. 381,968, on May 1, 1888, for an alternating current induction motor and the first U.S. patent, No. 382,280, on May 1, 1888, for the polyphase system of power transmission.

with being the first to invent a commercially successful ac induction motor and the three-phase alternating current to power the motor.

Nikola Tesla was born July 9, 1856, in Smiljan, Croatia, Province of Lika, then part of Austria-Hungary, (later part of Yugoslavia). His early education was conducted at elementary schools in Smiljan and Gospic and the secondary school in Karlovac. His father was an Orthodox Greek clergyman and planned for his son to follow in his footsteps. However, young Nikola wanted to be an engineer. In 1874, he was stricken with cholera and spent nine months in bed. To cheer him up, his father promised Nikola that he could study engineering. In 1875, he enrolled in the Polytechnic Institute at Gratz, Austria, and finished his studies at the University of Prague in 1880. It was while a student at Gratz that he became interested in ac motors after seeing a demonstration of Gramme's dc generator.

In 1881, while employed in the Central Telegraph Office of the Hungarian government at Budapest, Tesla made his first invention, a telephone repeater. It was also there at Budapest in 1882 that he discovered the principle of the rotating magnetic field. He moved to Paris that same year, and worked as an electrical engineer with the Continental Edison Company. He moved to Strasbourg the following year, and it was there in 1883 that he made the first models of the ac induction motor.

Wishing to go to the United States, he obtained from Charles Batchelor, an associate of Thomas Edison and an executive of the Continental Edison Company, a letter of introduction to Edison. He arrived in New York in 1884 with four cents in his pocket, having lost his wallet and all his spare clothes on the ship. As he walked up Broadway, he met a group of workmen trying to repair an electric motor. They paid him $20 to fix it.

Edison, on the basis of the Batchelor letter, gave him a job in his laboratory in Orange, New Jersey, designing motors and generators. The two men did not get along together, because Edison had committed himself to direct current while Tesla was devoted to alternating current. Tesla quit after Edison reneged on a promise he had made of a $50,000 bonus for improving the Edison generator.

With the financial help of A. K. Brown of the Western Union Telegraph Company, Tesla organized the Tesla Electric Company in April 1887, and began making his induction motors. On October 12, 1887, he filed a single patent application that covered his single and polyphase motors, his distribution system, and his polyphase transformers. The Patent Office required Tesla to break this single application into seven separate inventions. Within a short time, these seven applications matured into seven patents which were issued on May 1, 1888. The electrical engineering profession now took notice of this unknown inventor and invited him to deliver a lecture before the American Institute of Electrical Engineers on May 16, 1888. It was there that he delivered his classical lecture, "A New System of Alternate Current Motors and Transformers."

George Westinghouse recognized the commercial possibilities of Tesla's work and offered the inventor $1 million plus royalties for the exclusive right to his patents. Tesla quickly accepted and for about a year worked for the Westinghouse Company in Pittsburgh. In the spring of 1889, Tesla and Westinghouse put on the market a 1/6 horsepower alternating current motor that directly drove a three-blade fan. The battle between Edison's dc current and Westinghouse and Tesla's ac current was won by Westinghouse when this company was awarded the contract to build the first large-scale generating plant at Niagara Falls to supply the city of Buffalo with ac current. It was started in 1891 and completed in 1896, and used three Tesla ac dynamos of 5,000 horsepower each and Tesla's polyphase system of power transmission.

In 1890, Tesla returned to his laboratory in New York and from there in 1891 invented the "Tesla transformer," as well as many other electrical inventions. A fire destroyed his uninsured laboratory on March 13, 1895, but Tesla built a new laboratory with the financial help of J. Pierpont Morgan, John

Jacob Astor, John Hays Hammond, and Thomas Fortune Ryan. These investors expected him to make other great inventions, but he had few patents issued to him after 1905. Eventually, his backers either lost faith in him or died.

He became a paranoid egotist who experimented with oscillators, machines for making high-voltage sparks, wireless transmission of power, interplanetary communication, death rays, and a scheme for lighting the whole night side of the earth by artificial aurora. Although he conducted many famous demonstrations, nothing ever came from these experiments and demonstrations. He exhausted all his money and spent the last years of his life at the Hotel New Yorker living on a small pension from the Yugoslav government, with much of his time devoted to feeding pigeons. It was there that he died on January 7, 1943.

By the time of his death, both he and much of his work were forgotten. But while he was at the height of his creative endeavors, his name was always before the public. Fourteen universities presented him with honorary doctorates. Over 112 U.S. patents were issued to him. Over 108 magazines and periodicals and 28 newspapers carried more than 760 articles about him and his work. In 1912, the Nobel prize in physics was offered jointly to Tesla and Edison, but Tesla refused to be associated with Edison. The startled Nobel Committee gave the prize to Gustav Dalen, a Swedish inventor. He did receive the Edison Medal later after first turning it down.

Adding Machine

The first United States patent issued for a multiple order key-driven adding machine was No. 371,496 on October 11, 1887, to Dorr Eugene Felt.

Previous key-driven type machines, which were limited to the capacity of adding a single column of digits at a time, had been patented in the United States by DuBois Parmelee, No. 7,074, on February 5, 1850; David Carroll, No. 176,833, on May 2, 1876; and William Borland and Herman Hoffman, No.205,993, on July 16, 1878. However, it remained for Felt to perfect an adding machine on which multiple columns of digits could be added.

Dorr Eugene Felt was born on March 18, 1862, in Beloit, Wisconsin. After completing his schooling, Felt became employed as a machinist in a machine shop. In 1884, while operating a rachet-fed planing machine, he conceived the idea of making an adding machine with a rachet design. He knew that to be successful, the machine had to outperform accountants who could mentally add four columns of figures at a time.

After pondering the problem for a few months, he began gathering materials for his project. Lacking the necessary funds for constructing the machine out of metal, he decided to try his ideas with wood. He made a trip to the grocer's and selected a wooden macaroni box for the casing. From the butcher shop, he bought some meat skewers to be used for keys, and from the hardware, he bought staples for key guides and rubber bands to be used as springs. On Thanksgiving Day, in 1884, he started working on his adding machine with very few tools other than a jack knife. He soon discovered that it was necessary to make some parts out of metal. This delayed the completion of the machine until January 1885. In the fall of 1886, he finished a practical machine made entirely of metal.

Felt, then 24 years of age, started to manufacture his machine in the fall of 1886. Having little money, he had to make the machines himself. Under these conditions, he produced only eight finished machines before September 1887. He took one of the first eight machines to Washington and exhibited it

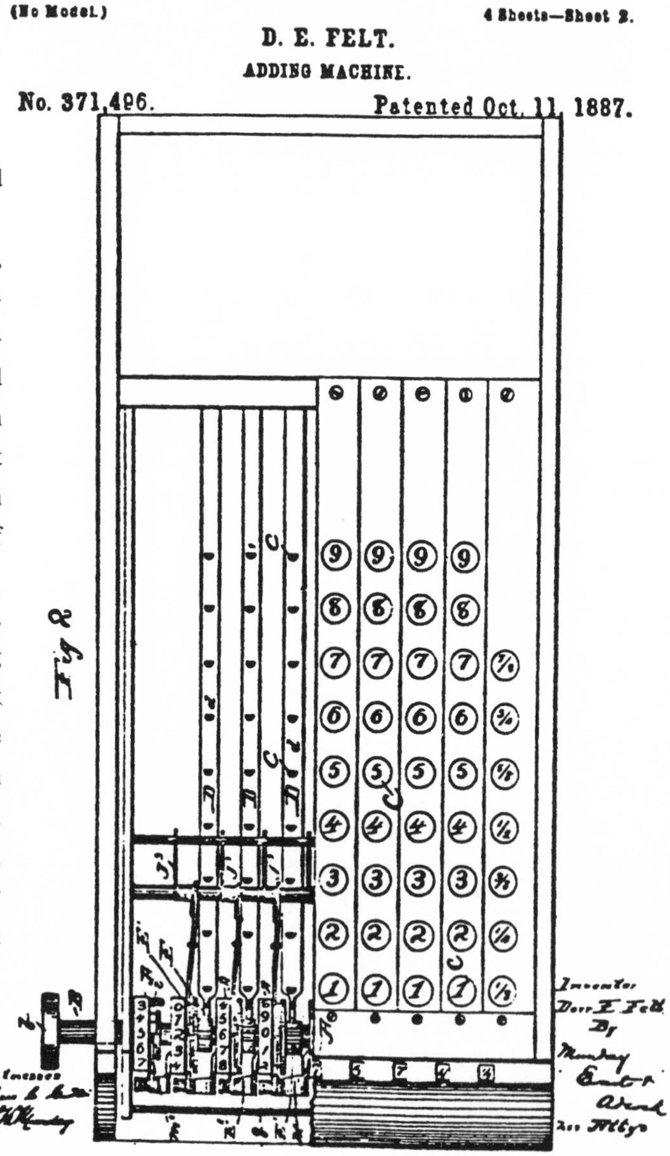

The multiple order key-driven adding machine, invented by Dorr Eugene Felt.

to General W. S. Rosecrans, Registrar of the Treasury. He was told to leave the machine in the office of Dr. E. B. Elliott, Actuary of the Treasury, where it was used constantly. Another of the first eight machines was placed with Dr. Daniel Draper of the New York State Weather Bureau in New York City. On November 28, 1887, Felt signed a partnership contract with Robert Tarrant of Chicago to manufacture his adding machines. They later incorporated under the name of Felt and Tarrant Manufacturing Company, which manufactured and sold the machines under the tradename Comptometer.

Felt's competition was William Seward Burroughs. While the Felt machine could add properly, it could not record and print the results. Burroughs constructed an adding machine that could accomplish this and received a patent in 1888. With this Adding and Listing Machine, he had a product that could successfully compete against Felt's. However, it was not long before Felt improved his machine by adding a recorder wherein a record of the operation was printed on a roll of paper. He filed for a patent on this improvement on January 19, 1888, which was issued June 11, 1889, as United States Patent No. 405,024. The Felt and Tarrant Manufacturing Company also made and sold this machine under the tradename Comptograph. The first machine of this type was placed on trial with the Merchants and Manufacturers National Bank of Pittsburgh, Pennsylvania, in December 1889, where it was used until 1899. This machine was later procured by Felt and presented to the National Museum of Washington, D.C.

Felt lived long enough to see his adding machines become a major part of the business world. He died April 7, 1930, in Chicago.

Addressograph

The first practical and manufactured addressing machine was the invention of Joseph Smith Duncan who received United States Patent No. 558,936 on April 28, 1896.

Other United States patents had previously issued on various forms of addressing machines but none had been practical nor reached the manufacturing stage. Some of the early patents were: H. Moeser, No. 8,175, issued June 24, 1851; S. D. Carpenter, No. 17,194, issued May 5, 1857; J. Lord, No. 21,429, issued September 7, 1858; G. Schuh, No. 23,787, issued April 26, 1859; and R. W. and D. Davis, No. 25,319, issued September 6, 1859. Duncan successfully solved the problems of previous workers and invented a practical addressing machine.

Joseph Smith Duncan was born near Pittsburgh, Pennsylvania, on April 5, 1858. His father, Joseph, was a farmer. Young Joseph, in addition to performing the necessary routine jobs on the farm, received his early education in the public schools of Allegheny County, Pennsylvania. His first job, at the age of 16, was at a local sawmill where his duty was to keep the logs fed to the saw. For this task, he was paid $4 weekly. It was there that he made his first invention, a device that caused the saw to stop when it had passed through the log. A short time later, he became a carpenter in the local community.

In 1879, he moved to Kansas where he started farming in addition to his carpenter work. His love for farming was no greater in Kansas than it was in Pennsylvania. He then moved to Sioux City, Iowa, and established a small construction firm to build bridges. This soon led to the establishment of a general contracting firm in Le Mars, Iowa. In 1885, he created a partnership with John Aupperle and organized a retail hardware store in Le Mars. The following year, Duncan was invited and accepted a position as manager of a bank in nearby Beresford, South Dakota. He observed that much of the bank's work consisted of addressing the names of depositors repeatedly on various statements. He wondered if a machine was available to do this monotonous task.

In 1887, Duncan returned to Sioux City as superintendent of a large flour and grist mill only to encounter the same problem of addressing by hand hundreds of cards to farmers to solicit business. He again inquired whether there wasn't some machine

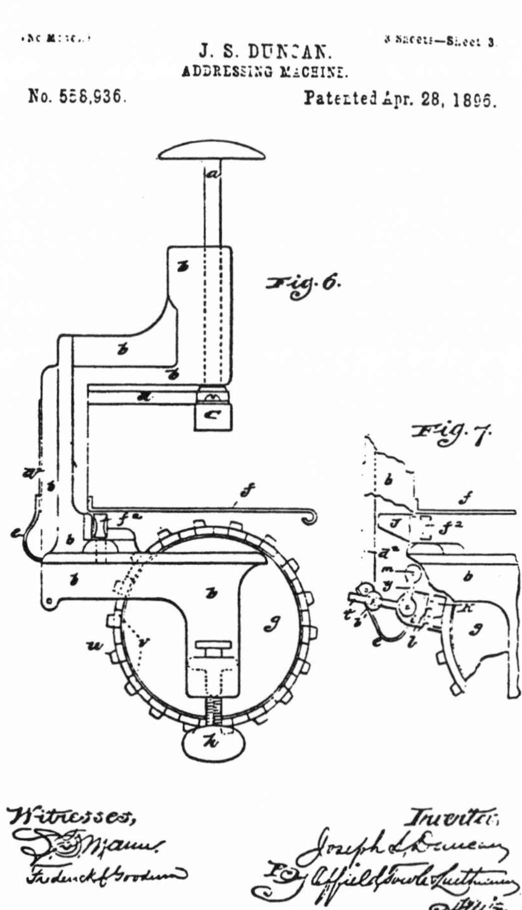

Joseph Smith Duncan's addressing machine, patented April 28, 1896.

to do this boring job. In fact, he even made a trip to Chicago to visit the leading stationery and office equipment stores but was told that no such machine existed. He returned to Sioux City determined to build for himself a machine that would perform this tedious and tiresome task.

Months of hard work followed. Finally in 1892, Duncan built his first crude addressing machine that he called the Addressograph. The first machine involved a rotating drum to which were affixed several names and addresses made from hexagonal wooden blocks. Glued to these were rubber type cut from rubber stamps. This machine was not particularly satisfactory, because the number of addresses was restricted to the capacity of the drum. Improvements followed with the result that Duncan designed and built what is now known as the first continuous chain addressograph.

The market for such machines wasn't good in Sioux City, so Duncan moved his base of operations to Chicago. In August of 1893, he ordered 25 machines from a Chicago manufacturer to be made under his supervision. The following January, the machines were ready for market. Within a short time, over half of the machines were sold. Enthused with his success, Duncan decided to expand his operation but lacked the necessary capital. Consequently, he interested John B. Hall, a Chicago businessman, in the project by selling him a half interest in the business for $5,000 and on January 1, 1896, Duncan and Hall formed a partnership and called it the Addressograph Company.

Many improvements followed, including the development of the Graphotype, a machine for embossing type into metal, and automatically operated machines, the first of which appeared in 1910 that ultimately reached speeds of 6,000 to 8,000 addresses an hour. In the fall of 1930, the Addressograph Company merged with the American Multigraph Sales Company to form the Addressograph-Multigraph Corporation with worldwide sales representation and distribution.

Duncan was issued more than 45 patents relating to the addressing machine. His company prospered to such an extent that he was able to give more than $1 million to various youth groups in Chicago, particularly the YMCA and different boys' clubs. He died in Los Angeles, California, on May 11, 1950.

Air Brake

The first commercially successful railroad braking system was the invention of the air brake by George Westinghouse Jr. For this invention, he received United States Patent No. 88,829 on April 13, 1869. The patent was later reissued as Reissue No. 5,504 on July 29, 1873. A reissue patent replaces the original patent to correct an accidental error, and expires on the date the original patent would have expired.

Westinghouse was not the only one interested in a braking system for large moving vehicles. Many other inventors did work in this area as well. The first United States patent for a wagon brake was granted to R. Turner of Ward, Massachusetts, on August 29, 1828. The first United States patent for a railroad brake was issued to Ephraim Morris of Bloomfield, New Jersey, on September 19, 1838. George Stephenson, a British railway engineer and inventor, in 1843, equipped one of his trains with a steam brake, but it was applied to only one car and not the whole train. Willard J. Nicholls placed a hand wheel on each car that operated a wooden block against the wheels. John Wilson of England was granted a patent for iron brake shoes. In 1846, Lucius Stebbins of Hartford, Connecticut, invented a brake that would act on all cars at the same time by means of a chain on a drum operated by the engineer. In 1860, Du Trembley received British Patent No. 1,737 of 1860 for a vacuum braking system. Even compressed air brakes were not new when Westinghouse perfected his invention with Wendell Wright having received United States Patent No. 12,263 on January 17, 1855 and Samuel Carson having received United States Patent No. 16,220 on December 16, 1856. However, it was George Westinghouse Jr. who invented the first practical and commercially successful application of air brakes to railroad equipment that resulted in the safe operation of modern railway systems.

George Westinghouse Jr. was born at Central Bridge, New York, on October 6, 1846. His father, George Sr., owned and operated G. Westinghouse and Company of Schenectady, New York, maker of agriculture equipment. Young George grew up and

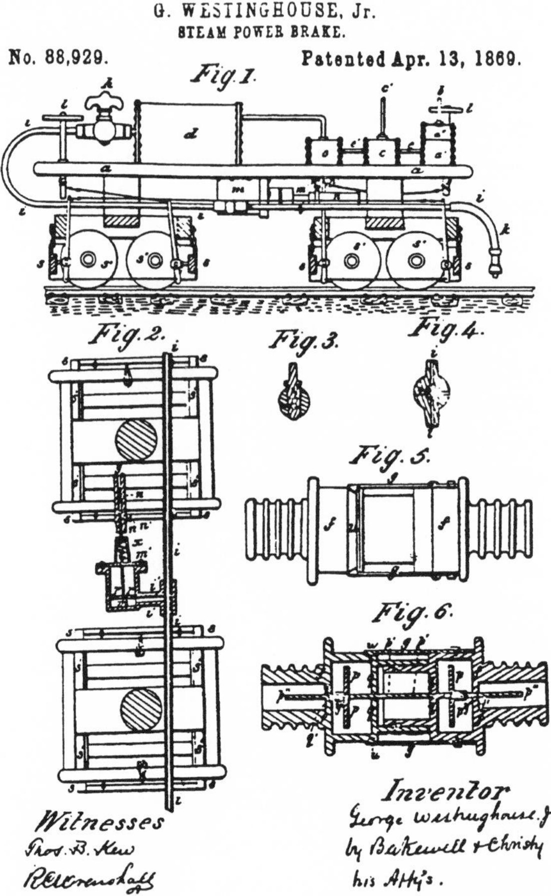

George Westinghouse Jr.'s air brake, patented April 13, 1869.

was educated in Schenectady and assisted his father in the shop. When the Civil War started, he followed his older brother and joined the Union Army at the age of 15. His father brought him back home but later permitted him to enter the cavalry at the age of 16. In

1864, he was honorably discharged and immediately joined the navy where, after passing an examination, he became acting third assistant engineer. He finished his navy career on the ships *Muscoota* and the *Stars and Stripes*.

In September 1865, at the insistence of his father, he entered Union College. Being more interested in machines than college work, and upon the advice of the college president, his father withdrew him at Christmas and put him to work in his shop for $2 a day. On October 31, 1865, while still in school, he obtained his first United States patent, No. 50,759, for a rotary steam engine. This was the first of some 361 United States patents he would receive. It came just over three months after his father had received United States Patent No. 48,857 for a sawmill machine, one of 30 that George Sr. received.

Train accidents played a large part in the life of George Westinghouse Jr. After one such accident, he spent many hours watching a train crew struggle to return two derailed cars to a track. Shortly thereafter, he invented a device to quickly do this task. At age 21, with the help of two Schenectady businessmen, he opened a factory, after receiving United States Patent No. 61,967 on February 12, 1867, and started manufacturing re-railers for railroads all over the country. Another accident occurred when two freight trains collided head-on. The engineers had seen each other in plenty of time but the brakemen, operating hand-wheels on each car, had not been able to stop the trains. Westinghouse believed that trains could be stopped more quickly if the engineer could apply the brakes throughout the train himself instead of depending on brakemen. After reading an article in the magazine *Living Age* on how the Mount Cenis Tunnel in Switzerland was built using air pressure for drilling from an air-compressor 3000 feet away, he knew that this was the power needed to operate his brake system. After perfecting the invention, he applied for a patent.

Westinghouse moved his re-railer plant to Pittsburgh and persuaded a rich young Pittsburgher to finance the air brake project and on September 28, 1869, the Westinghouse Air Brake Company was founded. He next persuaded the president of the Panhandle Railroad, W. W. Card, to let him test the brake. The test was completely successful. Later, Westinghouse equipped a train of 50 freight-cars with his air brake system and demonstrated it around the country, proving that his air brake could stop a train in one-fifth the distance than the old method of brakemen operating hand-wheels on each car. By 1876, over 38 percent of the trains in the United States were equipped with Westinghouse's air brake.

Westinghouse continued to improve his air brake. He invented and received United States Patent No. 175,886 on April 11, 1876, for an automatic brake system; No. 376,837 on January 24, 1888, for a quick acting air brake; and No. 448,827 on March 24, 1891, for a high speed air brake. He received more than 20 patents on the various aspects of the air brake.

The air brake was not the only contribution Westinghouse made to the industrial advancement of the United States. He was granted during his lifetime more than 400 foreign and domestic patents. His other interests included railroad signal systems, natural gas production, transmission and control, and alternating current electrical power transmission and utilization. He founded several large companies, including the Westinghouse Air Brake Company, the Union Switch & Signal Company, the Philadelphia Natural Gas Company, the Westinghouse Machine Company, and the Westinghouse Electric Company.

Many honors were given to Westinghouse. He was decorated by the French Republic and by the sovereigns of Italy and Belgium. He received the Grashof Medal, the highest honor in the engineering profession in Germany. He also received both the John Fritz Medal and the Edison Gold Medal.

George Westinghouse Jr. died of heart disease in New York City on March 12, 1914.

Air Conditioning

On a hot and humid day one can be thankful for the work of Willis Haviland Carrier, the father of air conditioning. For his efforts he received the first United States patent on modern day air conditioning. The patent, No. 808,897, was issued on January 2, 1906.

Air conditioning is a complex operation combining such features as refrigeration, humidifying, dehumidifying, purifying, and circulating. Previous to Carrier's patent, various individuals worked on these single operations. These included Jacob Perkins of Newburyport, Massachusetts, who in 1884 became the first in America to patent a refrigerating machine that used sulfuric ether in a closed cycle under compression. In 1838, David Boswell Reid of England added moisture to forced air to cool the House of Commons. In 1849, Dr. John Gorrie, an American physician from Charleston, South Carolina, invented an ice machine that was the forerunner of modern compressed refrigerating machines. In 1851, Ferdinand P. E. Carre of France designed the first ammonia absorption refrigerating machine. In 1856, Au Sieur Bandelot of France invented Bandelot coils for a refrigerating system. In 1856, Azel S. Lyman of New York discovered a crude method of air cooling by blowing air over ice in racks placed at the ceiling of a room. In 1866, Thaddeus L. C. Lowe, an American balloonist, developed a closed carbon dioxide refrigerating machine. In 1866, J. D. Whilpley and J. J. Storer of Boston, Massachusetts, invented an apparatus for removing dust from the air. In 1877, Franz Windhausen of Germany developed a water-vacuum refrigerating machine. In 1880, James G. Garland of Biddleford, Maine, invented an atomizing nozzle for supplying moisture to air. In 1884, Sir Oliver Lodge of England did research on the removal of dust from the air by electric precipitation. In 1882, Camille Edmond A. Rateau of France invented a multi-blade centrifugal fan for circulating air.

However, it was Carrier who combined these single operations and focused on the crucial relations between temperature and humidity. He invented modern day air conditioning.

Willis Haviland Carrier was born near Angola, New York, on November 26, 1876, on a farm that had been purchased years before by his great-grandparents. His father, Duane, was a postmaster before settling down to farming. Young Willis grew up as an only child with farm work and schooling consuming most of his time during his early years. In the fall of 1890, he entered Angola Academy and graduated

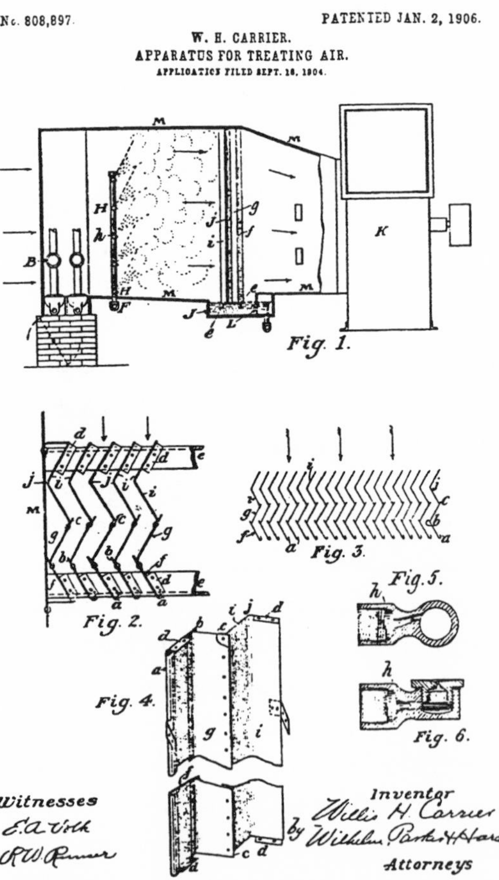

Willis Haviland Carrier's patent "Apparatus for Treating Air," granted January 2, 1906

four years later. His ambition was to enroll in Cornell University, but a national financial depression caused the farm to be mortgaged and forced Carrier to help out financially instead of going to college. He taught in one-room schools for the next two years.

Willis's mother died when he was 11 years old, and his father married Mrs. Eugenia Tifft Martin, a widow with three grown children, when Willis was 15. In 1896, his stepmother arranged for Willis to attend Central High School in Buffalo, New York, and to live with her son, a veterinarian in Buffalo. His ambition of going to Cornell University was realized the following year when he was awarded a state scholarship and the H. B. Lord Scholarship. He graduated from Cornell in 1901 with the degree of mechanical engineer in electrical engineering.

Carrier went to work for the Buffalo Forge Company on July 1, 1901, at a salary of $10 a week. In December of that year, he presented a paper entitled, "Mechanical Draft," based on work done after regular working hours, that so impressed the firm's top executives that they made him head of what was later an industrial laboratory. Carrier's first research was to obtain reliable information on how much heat air would absorb when air was circulated over steam-heating coils. When this work was completed, it was applied to the Buffalo Forge Company's installations and saved the company $40,000 the first heating season.

In the spring of 1902, Carrier was asked to solve a problem for the Sackett-Wilhelms Lithographing and Publishing Company of Brooklyn, New York. This firm printed such publications as the humor magazine *Judge*, whose covers were printed in color. During the summer, the paper stretched and shrank, due to humidity changes, so that the successive colors applied by the machines did not precisely register. Carrier solved this problem by devising a machine that controlled both temperature and humidity. On July 17, 1902, drawings were completed for the Sackett-Wilhelms system, which was later recognized as the world's first scientific air conditioning system.

Later, Carrier invented the conditioning equipment, for which he States Patent No. 808,987 on January designing this equipment, which not only the air but also dehumidified it, Carrier provided new industry that would accurately control an enclosed atmosphere. He also received many other patents on various improvements in the air conditioning field. He wrote several widely read books on the various aspects of air conditioning. His publication of the *Carrier Psychrometric Chart* brought him world-wide attention.

The demand for air conditioning systems was so great that in 1908 the Buffalo Forge Company established the Carrier Air Conditioning Company of America as a totally owned subsidiary to handle air conditioning sales, services, and engineering. By 1915, this company had installed over 500 central air conditioning systems in a wide variety of industries.

In 1915, the Buffalo Forge Company decided to cut back on its activities in the air conditioning business. This led to the founding of the Carrier Engineering Corporation of New York by Willis H. Carrier, J. Irvine Lyle, and five associates as an independent company. Carrier served as its president. The company was an immediate success and expanded its operation to include the European market.

In 1927, Carrier developed home air conditioning units and the following year established the Carrier-Lyle Corporation, a subsidiary of the Carrier Engineering Corporation, to handle the residential air conditioning business. In 1930, the Carrier Corporation of Newark, New Jersey, was formed through the merger of the Carrier Engineering Corporation and two other companies. Carrier was named Chairman of the Board and remained at this post until 1948 when he was made Chairman Emeritus.

After World War II, he cut back on his activities and took a three months tour with his wife. Everywhere they went, Carrier was honored by industrialists, scientists, and educators for his pioneering work in the air conditioning field. He died in New York City on October 7, 1950, of a heart ailment.

Airplane

l with mak-
g flight in
n machine.
ed the first
o. 821,393,

Many men before the Wright brothers had been intrigued with the idea of flying, principally with the aid of gliders. Even United States patents had been issued to such men as Otto Lilienthal, No. 544,816, on August 15, 1895; Octave Chanute, No. 582,718, on May 18, 1897; Pierre Mouillard, No. 582,757, on May 18, 1897; and John J. Montgomery, No. 831,173, on September 18, 1906. Other men who had received foreign patents or had written scientific articles were E. J. Marey, Clement Ader, Victor Tatin, William Henson, John Stringfellow, Samuel Pierpont Langley, and Hiram Maxim. However, it was the Wright brothers who succeeded in building, flying, and promoting the motor-driven heavier-than-air machine in controlled man-carrying flight, an invention that revolutionized travel.

Wilbur Wright was born at Millville, near New Castle, Indiana, April 16, 1867. Orville Wright was born at Dayton, Ohio, August 19, 1871. In all their enterprises the brothers were inseparable partners until the death of Wilbur due to typhoid fever on May 30, 1912. Their father, Milton, was a minister and bishop of the United Brethren Church and editor of *Religious Telescope*. Being a bishop caused the family to move frequently so that the boys attended public schools in Cedar Rapids, Iowa; Richmond, Indiana; and Dayton, Ohio. After entering high school in Dayton, the brothers became more interested in building mechanical devices than in structured classes and as a result never graduated. In fact, Wilbur and

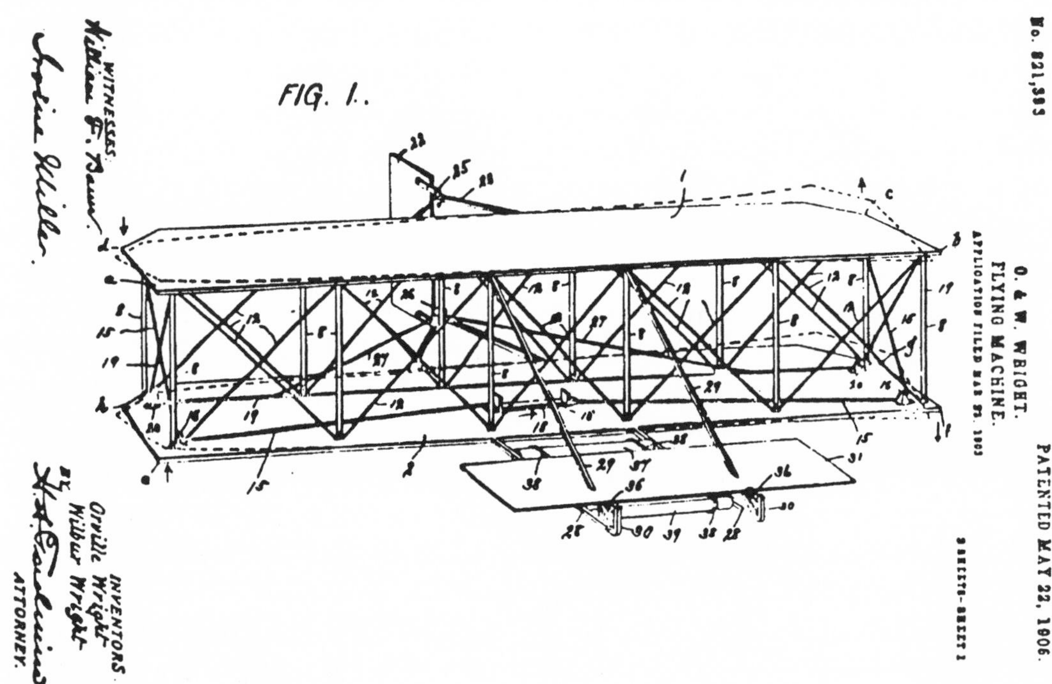

Orville and Wilbur Wright's Flying Machine, patent No. 821,383, May 22, 1906

Orville were the only members of the family who did not go to college. They spent their time building an eight-foot turning lathe and a treadle-operated paper-folding machine from wood and scrap. In 1888, they built a large printing press that turned out up to 1,500 copies an hour and from which they published a four-page weekly, called the *West Side News*. They did this for a few months until they realized it was more profitable to use their printing press for job printing.

In 1892, Orville, with his brother, opened a bicycle repair shop in Dayton, called the Wright Cycle Company, and by 1895 began constructing their own bicycles from purchased parts using their own designed and constructed tools. The shop was very successful and had to be moved several times to larger quarters. While conducting the bicycle shop, they became interested in the possibility of flying. They read all the literature they could obtain on the subject of aerodynamics, including the works of Langley, Mouillard, Maxim, Lilienthal, and Chanute.

After three years of study, they selected the Chanute biplane glider as the starting point and modified it by warping or twisting the trailing edges of flexible wings in opposite directions by means of a control cable, a method first suggested by Wilbur. This prevented the plane from tipping over to one side or the other in flight. This was the first scientifically designed wing. They also placed a rudder in front and suggested that the operator lay flat on the lower wing so wind resistance would be reduced.

From the United States Weather Bureau, they learned that at Kill Devil Hill, near Kitty Hawk, North Carolina, the wind blew smoothly and steadily. In 1900, they pitched a tent at Kill Devil Hill and began testing their new glider. They made nearly 1,000 glides and in many of them were able to stay in the air for over a minute. These tests convinced the Wright brothers that motorized mechanical flight was possible.

They returned to Dayton and built a wind tunnel in their bicycle shop from an 18-inch starch box with a glass top and tested more than 200 types of miniature wing surfaces in their tunnel. By 1902, they felt they had all the data necessary for building a practical airplane and that they were ready to make a trial with an engine. The gasoline engines at that time were too heavy for flight purposes, so they decided to build their own. This home-made four-cylinder engine weighed 179 pounds and developed 12 horsepower. They also developed the first formulas for propellers.

Returning to Kill Devil Hill in the fall of 1903, they took one of their best gliders and mounted their engine on the lower wing. The airplane, motor, and pilot weighed only 750 pounds. The plane had no wheels but had skids like sled runners. To launch it, a roller carriage supporting the plane was used which ran down a 60-foot inclined monorail. The power for increasing the speed of the plane-carrying carriage was supplied by a weight released from the top of a tower and connected to the carriage by a cable. After sufficient speed was acquired, the pilot tilted the horizontal rudder so the machine would rise from the carriage into the air.

Mishaps and bad weather postponed their flight until December 17, 1903. On this historic morning, Orville climbed into the machine, lay on his stomach, and gripped the controls. The engine was started and made to run a few minutes to warm up. The weight was dropped and then Orville released the wire that held the machine to the carriage. It shot forward into the wind and landed 120 feet from where it had risen into the air after a flight of 12 seconds. Later in the day, Wilbur took the controls and flew 852 feet in 59 seconds. After this flight, a gust of wind overturned and damaged the plane so badly that further flights were ended.

Besides the Wright brothers, only five other people witnessed this historic flight even though a general invitation had been issued to the community. These were John T. Daniels, W. S. Dough, and A. D. Etheridge from the Kill Devil Hill Life-Saving Station, W. C. Brinkley of Mantes, and Johnny Moore, a boy from Nags Head. Only a local newspaper carried an account of the flight and this was so inaccurate and so fantastic that people disregarded it. It was three years later in October 1905 that the *Scientific American* gave an accurate account.

The Wright brothers began a study of flight in 1897 and by 1903 were ready to file a patent application for a flying machine. They prepared and filed for patent protection in the Patent Office on March 23, 1903, without the aid of a patent attorney. The first office action by the Patent Office was received by the Wrights about a month later. In this letter, the drawings were found to be inadequate and the claims vague and indefinite, anticipated by six prior patents, and thought to be inoperative. The Wright brothers' response included a model to illustrate their wing warping control system. The examiner found the model of no help and again rejected the claims on the basis of inoperativeness. He further suggested that the applicants secure the services of a patent attorney.

After their historic flight on December 17, 1903, they decided to hire a patent attorney and on January 22, 1904, hired Harry A. Toulmin of Springfield, Ohio. Toulmin's first task was to file several foreign patent applications. Through amendments and interviews, Toulmin convinced the Patent Office that a patentable subject was present, and the application was issued as No. 821,393 on May 22, 1906. During the next few years, the patent was involved in numerous infringement suits but in each case, the court found the claims valid and infringed.

Although the first flight took place in 1903 and the patent was issued in 1906, it was not until 1908 that the public knew of their exploits when newspapers around the world reported about Wilbur's flights in France and Orville's flights at Fort Myer, Virginia.

After the death of Wilbur in 1912, Orville continued to work with airplanes as president of the Wright Company, founded in 1909. In 1915, he sold his entire interest in the company to a syndicate for more than $500,000. This syndicate later bought out Glen Curtiss, the Wright's greatest rival, and formed the Curtiss-Wright Company. In 1928, Orville loaned *The Flyer*, the plane that made history, to the Science Museum at South Kensington, England. In 1942, he asked the museum to return the plane to the United States after World War II upon settling his differences with the Smithsonian officials over which was the first man-carrying airplane in the history of the world capable of sustained free flight. Orville died January 30, 1948. His executors deposited the plane in the National Air Museum of the Smithsonian Institution on December 17, 1948.

Aluminum Manufacture

The first United States patent for a practical and commercial process for the manufacture of aluminum was awarded to Charles Martin Hall on April 2, 1889, as No. 400,664.

Other scientists who worked in this area before Hall included British chemist Sir Humphrey Davy, who was the first to prepare aluminum oxide in 1807 but could not separate the metal from its oxide. He named this metallic oxide alumina. The first aluminum metal was prepared by Hans Christian Oersted, a Danish chemist, in 1825 by heating aluminum chloride and potassium-mercury amalgam. In these experiments, the mercury evaporated leaving a lump of metallic aluminum. In 1845, Friedrich Wohler, a German chemist, isolated aluminum from its oxide and was the first to discover that it was an extremely light metal. Henri Etienne Sainte-Claire Deville, a French chemist, prepared the first bar of aluminum in 1855 by heating metallic sodium and aluminum chloride and exhibited it at the Paris Exposition. This metal captured the attention of Napoleon who ordered Deville to equip his army with aluminum helmets and breastplates. Deville built the first aluminum factory at the Salindres Foundry in Paris with money provided by Napoleon. However, it was Charles Martin Hall in 1886 who employed the process of direct electrolysis that has been used ever since for the production of metallic aluminum.

Charles Martin Hall was born at Thompson, Ohio, on December 6, 1863. His father, Herman Basset Hall, a minister, accepted a pastorate at Oberlin, Ohio, and moved his family there when Charles was quite young. His early interest in science was kindled by reading a chemistry book that his father had studied at Oberlin College. Since both his parents were Oberlin graduates, he enrolled in Oberlin after high school. While at Oberlin, Hall heard his professor of chemistry, Frank Fanning Jewett, state that if anyone could make aluminum cheaply on a commercial scale, he would become rich. This remark of Jewett's turned Hall's attention toward aluminum.

While still in college, Hall set up a laboratory in his father's woodshed. He was assisted by his next oldest sister, Julia, a former chemistry student at

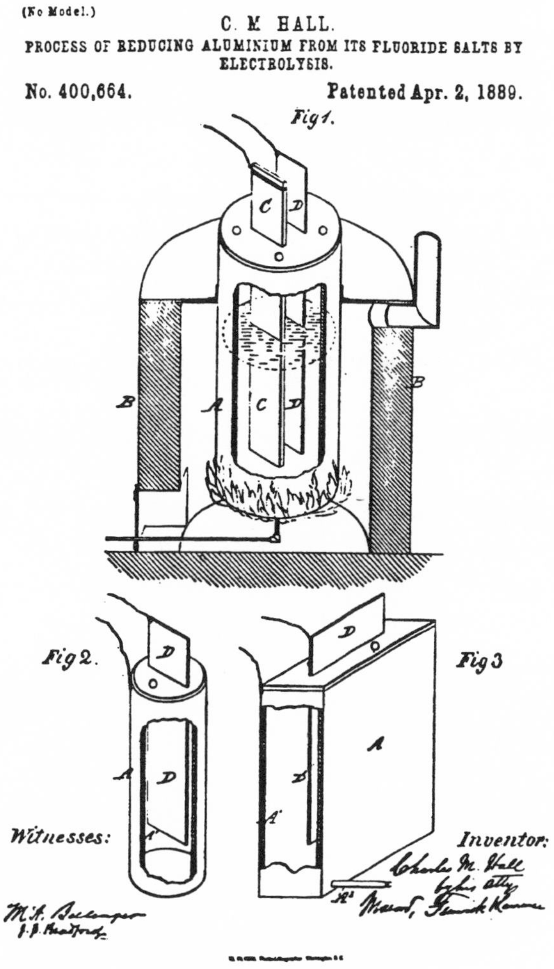

The commercial process for aluminum manufacture, Charles Martin Hall, April 2, 1889

Oberlin. After much experimentation, Hall discovered that the molten mineral cryolite would dissolve aluminum oxide and that this mixture could be electrolyzed. His first attempt at isolating aluminum by this process failed. He discovered that the failure was due to the clay crucible that held the cryolite-alumina mixture. He exchanged the clay crucible for one made of carbon and on February 23, 1886, made some small globules of aluminum. These first globules are still preserved by the Aluminum Company of America.

It is interesting to note that in 1886, Hall in America and Paul Louis Toussaint Heroult in France, who were both born in the year 1863 and who both died in 1914, independently and simultaneously discovered the electrolytic aluminum process. Both filed patent applications with the United States Patent Office, but Hall proved that he had completed his invention on February 23, 1886, while Heroult had to rely on April 23, 1886, the date he had filed his application in France. Today, it is usually called the Hall-Heroult process. The validity of Hall's patent was later sustained in an infringement suit brought by the Pittsburgh Reduction Company against the Cowles Electric Smelting and Aluminum Company in the United States Circuit Court for the central district of Ohio. The decision was rendered by Judge William Howard Taft on January 20, 1893.

After filing for his patent, Hall looked for financial backers and finally interested Captain Alfred E. Hunt and several Pittsburgh associates in forming the Pittsburgh Reduction Company in 1888. On Thanksgiving Day in 1888, the first aluminum ingot was poured at the New Kensington, Pennsylvania, plant. Needing more capital, the Pittsburgh Reduction Company attracted Andrew W. and Richard M. Mellon who gave the company a needed boost. The company changed the name to the Aluminum Company of America in 1907.

Hall's love for Oberlin College never diminished, and in 1905, he was elected as a trustee. In 1911, he was awarded the Perkin Medal in recognition of his services to the world. After his death on December 27, 1914, in Daytona Beach, Florida, he left to his alma mater a substantial part of his $30 million fortune.

Ammonia Process

Ammonia is an excellent source of nitrogen in compounds used in making explosives and fertilizers. Before Fritz Haber and co-worker Robert Le Rossignol synthesized ammonia from its elements, nitrogen and hydrogen, the large scale source for nitrogen compounds was in the nitrate deposits of northern Chile, a long way from most industrial sites. For this work, they received the first United States patent in this field as No. 971, 501 on September 27, 1910.

Previous to Haber, other workers had received United States patents for the preparation of ammonia, principally by absorption from illuminating gas and in the synthesis from saltpeter. These were: A. Paraf, No. 67,447, on August 6, 1867; L. S. Fales, No. 93,072, on July 27, 1869; J. J. Thomas, No. 178,889, on June 20, 1876; H. P. Lorenzen, No. 232,991, on October 5, 1880; and H. E. Baudouin and E. T. H. Delort, No. 454,108, on June 16, 1891.

Several attempts were made before Haber to produce ammonia on a large scale from its elements by passing them over a catalyst at a high temperature and pressure, but these were not commercially feasible. It remained for Haber, together with Le Rossignol, to find the optimum conditions and catalysts to produce ammonia from the atmosphere. Karl Bosch, a Badische Anilin-und Sodafabrik engineer, adapted Haber's method for commercial use. It is sometimes called the Haber-Bosch process.

Fritz Haber was born at Breslau, Germany (now Wroclaw, Poland), on December 9, 1868. His father, Siegfried, was a merchant of pigments and dyes. Young Haber attended the local elementary school and the St. Elizabeth Gymnasium in Breslau. He attended the Universities of Heidelberg, Berlin, Jena, and Karlsruhe and obtained his doctorate in organic chemistry in 1891 from the University of Berlin. He was self-taught in his chosen field, physical chemistry. He taught chemistry as a full professor at the Technische Hochschule at Karlsruhe from 1906 until 1911. When the Kaiser Wilhelm Institute for Chemistry was founded in Berlin in 1911, Haber was chosen as its first director. He remained there until 1933, and under his direction, this institution became one of the best chemical research centers in the world.

While at Karlsruhe, Haber started to investigate the possibility of synthesizing ammonia from its elements and on July 2, 1909, produced the first 100 grams of synthetic ammonia. He and his assistants accomplished this by mixing nitrogen, obtained from liquid air, with hydrogen in correct proportions and subjecting the mixture to high heat and pressure in

UNITED STATES PATENT OFFICE.

FRITZ HABER AND ROBERT LE ROSSIGNOL, OF KARLSRUHE, GERMANY, ASSIGNORS TO BADISCHE ANILIN & SODA FABRIK, OF LUDWIGSHAFEN-ON-THE-RHINE, GERMANY, A CORPORATION OF BADEN.

PRODUCTION OF AMMONIA.

971,501. Specification of Letters Patent. Patented Sept. 27, 1910.

No Drawing. Application filed August 13, 1909. Serial No. 512,679.

To all whom it may concern:

Be it known that we, FRITZ HABER, Ph. D., professor of chemistry, and ROBERT LE ROSSIGNOL, bachelor of science, subjects, respectively, of the King of Prussia and the King of England, residing at Karlsruhe, Germany, have invented new and useful Improvements in the Production of Ammonia, of which the following is a specification.

Several attempts have hitherto been made to produce ammonia on a large scale from its elements by passing them over a catalyst, but up to the present not much success has been met with.

In order that a process should be successful, it is advisable that the combination take place at as low a temperature and as quickly as possible, since when the temperature increases the concentration of the ammonia formed decreases.

We have now discovered that on passing gases containing nitrogen and hydrogen over osmium large quantities of ammonia can be obtained. This result is surprising, since it differs in this respect from the allied metal platinum (see *Zeitschrift für Elektrochemie*, vol. 14, p. 191).

In carrying out this invention, osmium can be used either in the form of the metal (preferably in a very finely divided condition) or in the form of a compound of the metal which upon being used becomes converted into metallic osmium, and the metal or its compound can be used either alone or in admixture with other substances or compounds. The osmium can be employed, for instance, in the form of metallic osmium, or it may be precipitated on a suitable carrier, such for instance as quartz, asbestos, clay, and the like. Asbestos containing ten per cent. of osmium is suitable for use. Further instead of metallic osmium, other suitable osmium compounds can be employed, such for instance as osmium oxid hydrate (prepared by the reaction of formaldehyde on an alcoholic solution of osmic acid, *cf. Berichte* 40, 1387), which under the action of the hydrogen used is converted into metallic osmium; or Fremy's salt can be used as the starting material, and either alone or mixed with an indifferent substance, or precipitated on a suitable carrier. Under the action of hydrogen it becomes converted into metallic osmium. The reaction can be carried out at ordinary pressure, but we prefer to carry it out under increased pressure, for instance at from 100 to 200 atmospheres.

As an example of the manner of carrying out the process of our invention, we give the following without in any way being confined to this example. Pass slowly a mixture of about three parts by volume of hydrogen and one part by volume of nitrogen over finely divided osmium at a pressure of one hundred and seventy-five atmospheres and at a temperature of about five hundred and fifty degrees centigrade. A yield of eight per cent. by volume of ammonia can easily be obtained.

Now what we claim is:

1. The process of producing ammonia by passing gases containing nitrogen and hydrogen over a catalyst containing osmium.

2. The process of producing ammonia by passing gases containing nitrogen and hydrogen over a heated catalyst containing osmium.

3. The process of producing ammonia by passing gases containing nitrogen and hydrogen under pressure over a heated catalyst containing osmium.

4. The process of producing ammonia by passing a mixture of nitrogen and hydrogen over a catalyst containing osmium at a pressure above 100 atmospheres.

5. The process of producing ammonia by passing a mixture of nitrogen and hydrogen over a heated catalyst containing osmium at a pressure above 100 atmospheres.

6. The process of producing ammonia by passing a mixture of hydrogen and nitrogen over heated osmium at a pressure above 100 atmospheres.

In testimony whereof we have hereunto set our hands in the presence of two subscribing witnesses.

FRITZ HABER.
ROBERT LE ROSSIGNOL.

Witnesses:
J. ALEC. LLOYD.
A. RENSLINGER.

This patent for the synthesis of ammonia from its elements was granted to Fritz Haber and Robert Le Rossignol on September 27, 1910

the presence of osmium as a metallic catalyst. The resulting ammonia gas was then liquified by cooling. This liquid ammonia could easily be transformed into explosives and fertilizers. For this work, Haber was awarded the 1919 Nobel Prize.

During World War I, Haber was a consultant to the German War Office. He and his associates developed a chlorine gas weapon that was used in 1915 against the Allies at Ypres, Belgium. After Germany's defeat, he attempted to recover gold from seawater so Germany might pay its reparations to the victorious Allies. After six years, the project was determined a failure.

In 1933, the Nazi anti-Jewish policy had no use for him because he was born a Jew. He was forced to resign as director of the research institute, because he refused to dismiss his Jewish staff members. He left Germany and accepted a position at the University of Cambridge in England. Shortly thereafter, Haber was offered a position at the Daniel Sieff Research Institute in Rehovot, Israel, by Chaim Weizmann, the first President of Israel. In route to Israel, he suffered a heart attack at Basel, Switzerland, and died January 29, 1934.

Anesthesia

A surgery patient under the influence of an anesthetic, as well as the medical profession, should be eternally grateful to a Boston dentist, William T. G. Morton, who, with Charles T. Jackson, received the first United States patent for an anesthetic. It was No. 4,848, issued November 12, 1846.

Before the discovery of an anesthetic, surgery was accompanied by excruciating pain. Many large hospitals located their operating rooms in an isolated and remote section of the hospital so the screams of the patients could not be heard by the other patients. Alcohol, opium, and hypnotism were tried as a means of decreasing the pain, but these had little effect. The first scientific attempt in the anesthetic field was in 1846 when Sir Humphrey Davy of England, inventor of the miner's safety lamp, used nitrous oxide or laughing gas for this purpose.

No previous United States patent had been issued on this subject. However, two men before Morton deserve mention. Crawford W. Long, a physician of Jefferson, Georgia, used sulfuric ether as an anesthetic in the removal of a cystic tumor on March 30, 1842. He used ether in other operations including the delivery of his second child, Fanny. After being threatened by the townspeople, he didn't make his experiments known to the medical profession until three years after Morton had applied for a patent. Horace Wells, a Connecticut dentist and a business associate of Morton, used nitrous oxide as early as 1844 for the extraction of teeth, after first testing its effectiveness on himself. Convinced that he had discovered something of great value, he arranged for a demonstration before the medical class of Harvard College at the Massachusetts General Hospital. The test was a failure. Humiliated, Wells tortured himself over his failure and finally went insane after becoming an ether addict. He later committed suicide while under the influence of ether.

William Thomas Green Morton was born August 9, 1819, at Charlton, Massachusetts. His father, James, was a local farmer. After receiving a common school education at Northfield and Leicester Academies, he went to Baltimore in 1840 and began the study of dentistry at the College of Dental Surgery, the first dental school in the United States. After two years of study, he began a dental practice in Farmington, Connecticut. In 1842, he met Horace Wells, a dentist in Hartford, Connecticut, who later, in 1844, used nitrous oxide in the extraction of teeth. They formed a partnership, moved to Boston and practiced together until the autumn of 1843 when

UNITED STATES PATENT OFFICE.

C. T. JACKSON AND WM. T. G. MORTON, OF BOSTON, MASSACHUSETTS; SAID C. T. JACKSON ASSIGNOR TO WM. T. G. MORTON.

IMPROVEMENT IN SURGICAL OPERATIONS.

Specification forming part of Letters Patent No. 4,848, dated November [illegible]

The first U.S. patent for an anesthetic was issued to William T. G. Morton and Charles T. Jackson on November 12, 1846.

Wells moved back to Hartford. Morton remained in Boston, and enrolled at Harvard Medical School but due to financial problems never received a medical degree.

In 1844, Morton met Charles T. Jackson, a renowned chemist and ex-physician. Jackson suggested to Morton that he use sulfuric ether for his dental work instead of the nitrous oxide that had been used by Wells. On September 30, 1846, Morton used ether, previously obtained from Jackson, to extract a tooth. The Boston Daily Journal reported this event to the public on October 1, 1846. This announcement eventually reached the attention of Henry Jacob Bigelow and John Collins Warren of the Massachusetts General Hospital. Arrangements were made for Morton to try his discovery at Massachusetts General. On October 16, 1846, Warren removed a tumor from the neck of a patient named Gilbert Abbott, with Morton administering the ether. The operation was a complete success, with the patient showing no evidence of pain. The operation also created a sensation in the medical world. Oliver Wendell Holmes, the famous physician and author, suggested the name "anesthesia" to Morton.

After the demonstration, Jackson demanded $500 from Morton for his advice and chemicals. Morton did not have the money but agreed that the patent should be taken out in both their names and that Morton should pay Jackson 10 percent of the earnings from the invention. Morton had little success in collecting royalties on his patent. In 1847, the French Academy of Medicine awarded the Monthyon prize of 5,000 francs to Jackson and Morton jointly, but Morton refused because he believed that the invention was his alone. He applied to Congress for aid, and a bill proposing a grant of $100,000 was submitted in 1852. This grant might have passed if the supporters of Jackson, Wells, and Long hadn't come forward with testimony that each of these men was the real inventor of anesthesia.

The patent to Morton and Jackson was held valid by the courts several times until Morton sued The New York Eye Infirmary for infringement. The court held in that case that the application of ether to surgical purposes resulting in anesthesia was a discovery and not an invention and as such was not patentable.

After serving as a volunteer anesthetist in the Union Army in the Civil War, Morton retired to a farm, poor and bitter. On July 15, 1868, he died of a stroke while riding in a carriage in Central Park, New York. It was reported the stroke was brought on when Morton read a paper that stated Jackson was the inventor of anesthesia. Morton was later elected to the Hall of Fame for Great Americans in 1920.

While some credit for the development of an anesthetic may be given to Long, Wells, and Jackson, it was William T. G. Morton who publicly demonstrated its usefulness on October 16, 1846, and received, together with Jackson, the first United States patent for an anesthetic on November 12, 1846.

Artificial Limb

The first United States patent for an artificial limb, No. 4,834, was granted to Benjamin Frank Palmer on November 4, 1846.

Palmer was by no means the first to substitute a mechanical device for the loss of an extremity. The first artificial limb of known record was described by Greek historian Herodotus and belonged to a man named Hegesistratus, who was imprisoned and condemned to die in 484 B.C. Hegisistratus had cut off his foot to escape his captors, and, after healing, made for himself a wooden foot. He was later recaptured and put to death by the Spartans. The best known and oldest artificial limb in existence was found in an ancient tomb in 1858 in Capua, Italy, and is on display at the Royal College of Surgeons in London. It is made of wood and reinforced with bronze, leather, and iron and dates back to the Samnite Wars of 300 B.C.

In other accounts, a Roman general, Marcus Sergius, lost his right hand during the second Punic War of 210 B.C., and was fitted with an iron hand, according to the writings of Pliny, the elder, in 61 A.D. There is very little history on artificial limbs after this until 1400 when the Alt-Ruppin hand was found in diggings along the banks of the Rhine. In this hand, the thumb was rigid and the fingers were

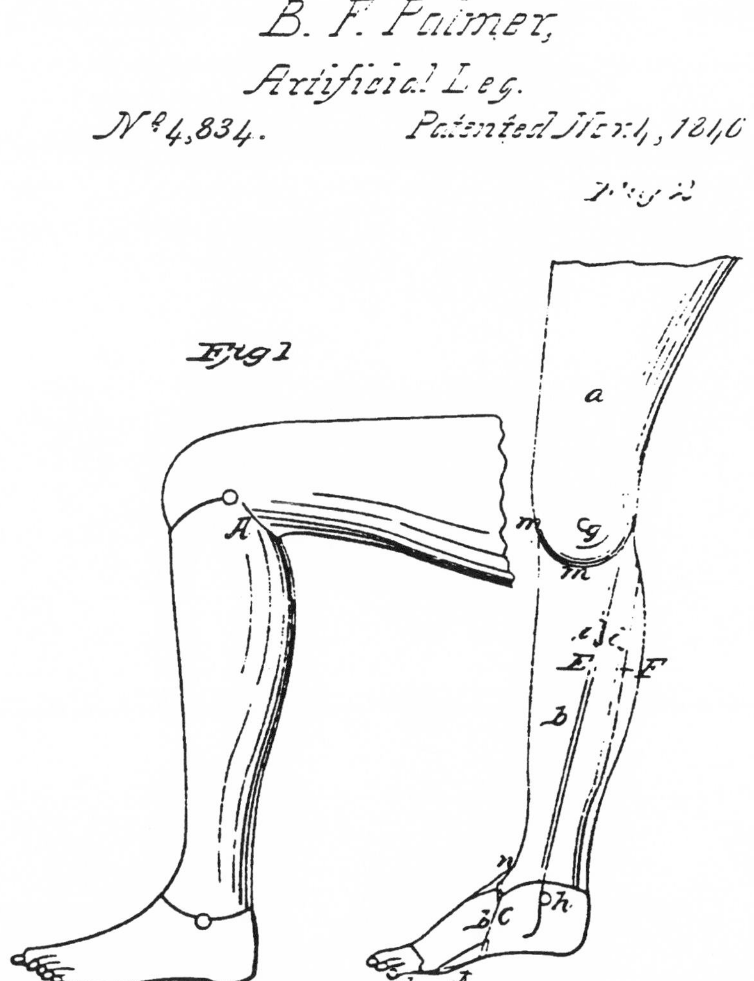

moved in pairs by a spring mechanism attached to buttons at the base of the palm. An artificial hand was also made in 1509 for Goetz von Belichinger, a German knight, that was equipped with jointed fingers and is now preserved in the Nurnberg Museum.

Ambrose Pare, chief surgeon to Charles IX of France and founder of modern principles of amputation, had a locksmith, whom he called le petit Lorrain, construct artificial legs of iron and leather with moveable ankle and knee joints in 1561. Verduin, a Dutch surgeon, in 1696 invented the first artificial leg with an adjustable socket so the size of the socket could be reduced to meet the needs of a shrinking stump. In 1775, Ravaton constructed an artificial leg that consisted of a laced leather boot filled with horsehair and that had an ankle joint that could be controlled by a spiral spring. Others who contributed to the art were Charles White, J. H. Brunninghousen, Gavin Wilson, Johann Georg von Heine, and Peter Ballif. Ballif, a German dentist, appears to have been the first to introduce the use of the trunk and shoulder muscles as sources of power to flex or extend fingers in 1818.

On November 15, 1800, James Potts of England patented an artificial leg made of two hollow wooden cones with a steel knee joint and a wooden ankle joint that was controlled with cords from the knee joint. One of these legs was made for the Marquis of Anglesey, who lost his leg in the Battle of Waterloo, and became known as the Anglesey leg. William Selpho, an employee of Potts, introduced this leg in the United States in 1839, and became one of the first manufacturers of artificial legs in America. Selpho made an artificial leg for Benjamin Frank Palmer, who then modified the leg and filed an application for a patent. The patent issued on November 4, 1846, and was the first United States patent on artificial limbs issued by the United States Patent Office.

Little is known about the life of Benjamin Frank Palmer except that he was an amputee and lived in Meredith, New Hampshire, when he received his patent. However, Palmer's artificial leg did create considerable interest in the United States. There were nearly 250 patents issued for artificial limbs in the United States between 1846 and 1895.

Atomic Reactor

The first atomic or nuclear reactor was made by Enrico Fermi, who, in cooperation with others, was able to not only produce but also control the energy released by neutron bombarded atoms of uranium in a chain reaction. Fermi, together with Leo Szilard, was granted the first United States patent, No. 2,708,656, for an atomic reactor on May 17, 1955.

Like many practical scientific inventions, Fermi relied heavily on the academic and theoretical work of previous investigators. These would certainly include the works of Antoine Henri Becquerel of France who discovered radio-activity in 1896, James Chadwick of England who discovered the neutron in 1932, Frederic and Irene Joliot-Curie of France who discovered artificial radio-activity in 1934, Otto Hahn and Fritz Strassmann of Germany who achieved fission of the element uranium by bombarding it with neutrons in 1938, and Otto Frisch and Lise Meitner of Austria who recognized that an immense release of energy takes place with these fissions in 1939. However, it was Fermi who was responsible for the control of this vast amount of energy.

The patent granted to Fermi and Szilard was assigned to the United States of America as represented by the United States Atomic Energy Commission. Five years later, the first atomic reactor system was patented by a private company, North American Aviation, Inc. The patentee was John William Flora of Canoga Park, California, who received United States Patent No. 2,937,127 on May 17, 1960.

Enrico Fermi was born in Rome, Italy, on September 29, 1901. His father, Alberto, was a railroad official. While still a teenager, young Fermi independently through study and with the help of an engineer

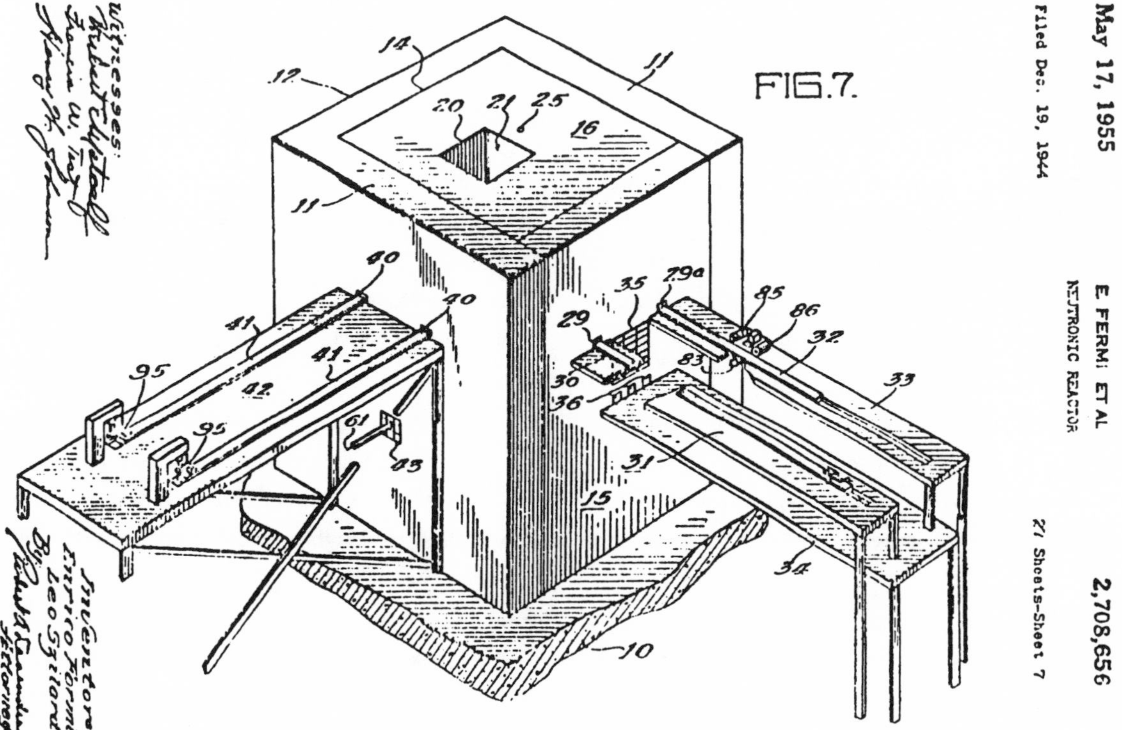

The first U.S. patent for an atomic reactor, granted to Enrico Fermi and Leo Szilard on May 17, 1955

friend gained an advanced knowledge of mathematics and physics. This knowledge won him a fellowship in 1918 to the Scuola Normale Superiore of the University of Pisa. Four years later, he received his doctoral degree in physics magna cum laude from the University of Pisa. He did post-graduate work at the University of Gottinger under Max Born and the University of Leiden under Paul Ehrenfest. In 1924, Fermi became a lecturer of mathematical physics and mechanics at the University of Florence. His next teaching assignment was professor of physics at the University of Rome, where, for the next few years, he studied the creation of artificial radioactive isotopes through neutron bombardment. For his work in this field, he was awarded the Nobel Prize in physics in 1938.

It was during this period that times became intolerable for Fermi and his family. Hitler's influence had become more pronounced in Italy with the passing of anti-Jewish laws. With his wife being Jewish and Fermi himself being anti-Fascist, their popularity in Italy was at a low ebb. Thus, Fermi seized the opportunity, while receiving the Nobel Prize in Stockholm, to leave Italy and come to the United States as professor of physics at Columbia University.

Shortly thereafter, Fermi, in cooperation with Leo Szilard, graduate student Herbert Anderson, and researcher Walter Zinn, began experiments on nuclear fission. They determined that enough neutrons were released during fission for the possibility of a chain reaction. In March, 1939, George Pegram of Columbia University arranged for Fermi to discuss the matter with the Navy. The Navy appeared interested, but nothing came of the meeting. In July, 1939, Leo Szilard explained to Albert Einstein the nature of the work, and Einstein, in turn, sent a communication to President Roosevelt. A meeting was subsequently arranged between several scientists and officials of the armed services.

In November, 1940, $40,000 was assigned to Columbia University for research on the materials and to provide conditions necessary for a controlled nuclear fission by a research team led by Fermi. Their work was checked by a Princeton University team under the direction of S. K. Allison. In January 1942, the Columbia and Princeton teams combined forces and moved to Chicago. On December 2, 1942, the first controlled chain reaction of nuclear fission was put into operation on the floor of a squash court under the west stands of Stagg Field at Chicago University.

In July, 1944, Fermi and his family became naturalized United States citizens and the following month moved to Los Alamos, New Mexico, where he became chief of an advanced physics department of a new laboratory, headed by J. Robert Oppenheimer. The mission of this new laboratory was to build an atomic bomb.

One year and $2 billion later in a remote section of the Alamogardo Air Base, 120 miles southeast of Albuquerque, New Mexico, the first man-made atomic explosion was witnessed at 5:30 a.m. on July 16, 1945, by a group of renowned scientists and military men. The public first heard about the explosion three weeks later on August 6, 1945, when the towns of Hiroshima and Nagasaki in Japan were destroyed by two atomic bombs. While these atomic bomb explosions were important in the history of modern warfare, the day to be remembered is December 2, 1942, when a group of scientists working under the direction of Enrico Fermi, showed that nuclear fission could be controlled and made to yield useful power for the benefit of mankind.

After World War II, Fermi returned to the University of Chicago to become the Swift Distinguished Professor of Physics. He died of stomach cancer in Chicago on November 30, 1954. The following year, the newly discovered element 100 was named fermium in his honor.

Fermi received many honors. In addition to the Nobel Prize in physics, he was awarded the Matteucci Gold Medal of the National Academy of Sciences of Italy in 1926, the Hughes Medal of the Royal Society of London in 1943, the Civilian Medal of Merit of the United States government in 1946, the Franklin Medal of the Franklin Institute in 1947, the Bernard Gold Medal for Meritorious Service to Science of Columbia University in 1950, and the first Fermi Prize of the United States Atomic Energy Commission in 1954.

Automobile

The Selden automobile was never manufactured even though the first United States patent for an automobile was awarded to George B. Selden on an application filed May 8, 1879. Through a series of legal delays, the patent, No. 549,160, was not issued until November 5, 1895.

Many people contributed to the development of the automobile, starting with steam as the source of power. In 1678, Ferdinand Verbiest, a Jesuit missionary in China, built a working model of a steam-powered automobile. In 1769, Nicolas Joseph Cugnot, a French Army officer, built the first steam self-propelled vehicle for military purposes. In 1787, American inventor Oliver Evans obtained a patent in Maryland for the exclusive right to make steam road wagons. He built the first American self-propelled vehicle, a 40,000-pound steam-powered vehicle that could operate on both land and water. On August 26, 1791, Nathan Reed received a United States patent for a steam carriage. Others followed, including Richard Trevithick, Sir Goldworthy Gurney, William Church and Richard Dudgeon, but perhaps the most famous was the Stanley Steamer by Francis E. and Freelan O. Stanley, who received United States Patent No. 657,711 on September 11, 1900.

The invention of the internal combustion gasoline engine was a prerequisite for the Selden automobile. United States patents had been issued to Samuel Morey on April 1, 1826, for an internal combustion engine; to George B. Brayton on April 2, 1872, No. 125,166, for a two-cycle gasoline engine; and to Nicolaus A. Otto on August 14, 1877, No. 194,047, for a four-cycle gasoline engine.

However, Selden was the first to combine an internal combustion engine with a carriage and to file an application for a United States patent. The automobile that he disclosed had, in addition to an engine, a clutch, foot-brake, muffler, front-wheel drive, and power shaft arranged to run faster than the propelling wheel.

George Baldwin Selden was born in Clarkson, New York, on September 14, 1846. His father, Henry, was lieutenant governor of New York in 1856 and

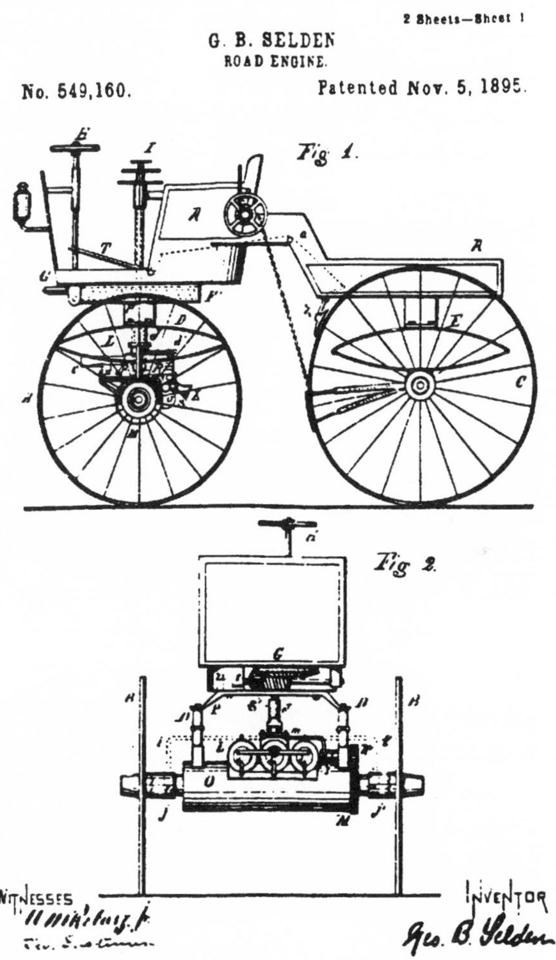

G. B. Selden's Road Engine, patented November 5, 1895

later a judge of the Court of Appeals. Young Selden attended the local schools and St. Albans Classical Preparatory School. He then attended the University of Rochester but did not finish before enlisting in the 6th New York Cavalry at the beginning of the Civil

War. In 1865, Selden enrolled in Yale College to study classics but later changed to the Yale Sheffield Scientific School to study engineering. His father's illness forced him not only to leave school, but to give in to his father's wish that he study law. For three years, he was an apprentice in his father's law office, where he became versed in patent law. Then in 1871, Selden was admitted to the bar. He later established his own patent practice and handled the application for George Eastman's initial patent on an improved process for coating photographic plates, a process that led to the formation of Eastman Kodak Company. Later, George Eastman was one of the witnesses who signed the patent drawings for Selden's automobile patent.

Whenever his law practice permitted, Selden tinkered with inventions in a basement workshop. Some inventions for which he received patents were a machine to manufacture barrel hoops, a device for attaching solid rubber tires to wheels, and a typewriter. However, his main interest was to invent a self-propelled road vehicle. After many years of study, he decided that the solution lay in reduced weight and increased power.

Visiting the Philadelphia Centennial Exposition in 1876, Selden studied an array of internal combustion engines and concluded that the Brayton engine, which used crude petroleum, would most satisfactorily supply the power needed to drive his road vehicle. He redesigned this engine by enclosing the crankcase and in 1878 produced an engine weighing 370 pounds that developed two horsepower. After further experimentation, he filed a patent application on May 8, 1879, with the United States Patent Office on his road vehicle.

In those days an applicant had two years in which to answer an office action. By taking the full two years to answer each office action, Selden prevented his application from being issued until some 16 years had passed. This permitted the automobile industry, practically non-existent at the time Selden filed, to catch up with his invention, giving Selden an opportunity to collect royalties on each automobile manufactured for 17 years after the patent was granted.

Failing to find financial backers, Selden in 1899 sold the rights to his patent to William C. Whitney of the Columbia Motor and Electric Vehicle Company of New Jersey for $10,000, retaining only certain royalty rights. In 1900, this company successfully sued the Winton Motor Carriage Company for infringement. As a result 10 automobile makers banded together to form the Association of Licensed Automobile Manufacturers and purchased the right to use the Selden patent at a royalty of 1 1/4 percent of the retail price of all automobiles sold. During this agreement, which lasted about eight years, the Association and the Columbia Motor Company received more than $2 million each and Selden was paid about $200,000 as his share of the royalties.

Henry Ford applied for a license under the Selden patent but was refused on the basis that he was "an assembler and not a builder of cars." Ford disregarded the refusal and continued to build his cars. In 1903, Selden and the Columbia Motor Company brought an infringement suit against the Ford Motor Company, which had refused to pay any royalties. Ford fought the suit all the way to the United States Circuit Court of Appeals for the Second Circuit, where on January 11, 1911, a decision was handed down stating that the Selden patent was valid but that Ford was not guilty of infringement because he was using the Otto four-cycle engine while the patent was limited to the Brayton two-cycle engine. This decision had far-reaching effects on the automobile industry. Because all the automobile makers were now using the Otto type engine, it was no longer necessary to pay Selden any royalties for the manufacture of automobiles.

Selden retained his patent law practice and tried unsuccessfully to get into the automobile manufacturing business. He died in Rochester on January 17, 1922.

Bakelite

The modern plastics industry had its beginning with Belgian-American chemist Leo Hendrik Baekeland, who received the first United States patent for a thermo-setting plastic on December 7, 1909. The patent was No. 942,809.

Previously, other workers had struggled with the reaction of phenol and formaldehyde to produce a synthetic resin, but without success. All that was produced was a black unworkable substance full of holes. It remained for Baekeland to correct the mistakes of his predecessors by increasing the temperature and pressure and using alkali instead of acid to control the reaction. The result was the first thermo-setting resin, one that once set would not soften under heat.

Leo Hendrik Baekeland was born in Gent, Belgium, on November 14, 1863. He attended the public schools of Gent and graduated from high school at the age of 16. He then attended the University of Gent on a scholarship and in 1884 obtained his doctorate at the age of 21. After obtaining his doctor's degree, he was appointed professor of chemistry and physics at the Government Higher Normal School at Bruges, Belgium. Following that, he was appointed as associate professor at the University of Gent.

After a short time at the university, Baekeland won a three-year traveling fellowship and in 1889 arrived in New York. While attending a meeting of the New York Camera Club, he met and impressed Richard Anthony of the A & H T Anthony Company, the largest photographic firm in the city. Anthony made Baekeland an offer to be a chemist in the company. After resigning from the University of Gent, he joined the Anthony Company where he remained for two years.

Baekeland left Anthony in 1891 to become a consulting research chemist. It was during this period that he perfected the process for the manufacture of Velox, a photographic paper sufficiently sensitive for printing by artificial light. He met a retired stock broker, Leonard Jacobi, who offered to financially back him in the manufacture of his photographic paper. They established the Napera Chemical Company at Napera Park, Yonkers, New York, to manufacture the paper. It took six years to convince the picture-taking public that Velox was a good product. After that, the company prospered so much that George Eastman of the Eastman Kodak Company offered Baekeland $1 million for his interest in the enterprise which he accepted. Baekeland said later

UNITED STATES PATENT OFFICE.

LEO H. BAEKELAND, OF YONKERS, NEW YORK.

CONDENSATION PRODUCT AND METHOD OF MAKING SAME.

942,809. Specification of Letters Patent. Patented Dec. 7, 1909.

No Drawing. Application filed October 15, 1907, Serial No. 397,560. Renewed September 17, 1909. Serial No. 518,283.

To all whom it may concern:

Be it known that I, LEO H. BAEKELAND, a citizen of the United States, residing at Yonkers, in the county of Westchester and State of New York, have invented certain new and useful Improvements in Condensation Products and Method of Making Same, of which the following is a specification.

This invention relates to an improved method of reacting with formaldehyde upon phenol or a phenolic body, and the improved product resulting from such reaction.

The condensation products resulting from the chemical action of aldehydes on phenols have received by further treatment some industrial applications in the manufacture of varnishes, resinous products and plastic compounds. In some cases such condensation products have been prepared by simple boiling or heating of phenol and formaldehyde without the addition of condensing agents; but such treatment does not in all cases yield the desired result, and at best the reaction is very slow, requiring about eight hours boiling with ordinary commercial phenol and formaldehyde. With pure crystallized phenol the reaction does not occur even after forty-eight hours constant boiling. It has also been proposed to use acids or salts as condensing agents, but the employment of these results in a stormy reaction, often difficult to control, and yields products containing undesirable impurities which it is difficult or impracticable to eliminate, these impurities having the effect of causing the resulting mass to darken with age or in presence of alkalies.

I have discovered that the addition in proper proportions of an organic or inorganic base to a mixture of phenol and formaldehyde, or to either component of the mixture, facilitates the reaction and yields products which are commercially far superior to those obtained by simple heating or by the use of acids or salts as condensing agents. The proportion of phenol to formaldehyde may be considerably varied, the formaldehyde being present in about the molecular proportion required in the reaction or in excess thereof. The base may be added at any phase of the process, either at the start or during the heating of the mixture, or in successive portions as the heating proceeds.

According to my invention the alkalies or bases are used in such relatively small proportions that their presence does not interfere with the desirable qualities of the products, rendering it unnecessary to eliminate them by washing or neutralizing. In fact in most cases the small amount of base persists in the final products and confers upon them new and desirable properties.

By reason of their comparative cheapness it is preferred to employ as a condensing agent ammonia, anhydrous or aqueous, ammonium carbonate, caustic alkalies or their carbonates, anilin or pyridin, but other bases as for instance the hydrates of barium, strontium, or calcium may be used. Amins and amids, and in general all derivatives of the type NH_3 which possess basic properties are found to act in the same manner. Similarly all basic salts, or salts which by secondary reaction engender bases, as for instance alkali sulfids, acetates and cyanids, sodium triphosphate, borax, soaps, etc., may be used; also alkali sulfites may serve, for the reason that when boiled with formaldehyde they liberate alkali in accordance with the well known reaction:

$$CH_2O+SO_3Na_2+H_2O=CH_2(OH)SO_3Na+NaOH$$

The bases above referred to, and others having the requisite basic properties, are employed in variable proportions, according to their character and also according to the result desired. Additions of ammonia or caustic soda in so small a proportion as one-half per cent. of the weight of phenol used show a decided influence, but in most cases it is desirable to use somewhat larger proportions, rarely attaining however 10 % by weight of the phenol or phenolic body. The proportion of bases used as condensing agents has a preponderant influence on the nature of the ultimate products. For instance, if a large amount of ammonia be used, hexamethylentetramin is formed, which is a crystalline body of definite chemical properties. (See Wohl, *Ber.*, 19, 1892; Tollens, *Ber.*, 17, 653. See also Moschatos and Tollens *Ann. der Chemie*, 272, 280.) Likewise, if large amounts of caustic soda be used there are obtained alkaline derivates of phenol-alcohol. (See Lederer, *Journal Praktische Chemie* (2), Vol. 50, page 224,

The first United States patent for a thermo-setting plastic was granted to Leo Hendrik Baekeland on December 7, 1909

that he had planned to ask $50,000 and go down to $25,000 if necessary, but fortunately Eastman spoke first.

With this newly acquired capital, Baekeland built for himself a private laboratory. He devoted the next few years to studying electrochemistry, out of which came a process for manufacturing caustic soda and chlorine, which helped to perfect the Townsend process at the Hooker plant in Niagara Falls, New York.

Baekeland fell ill and while recuperating decided to pursue only one of the many ideas that was running through his mind. He began the work that would bring him fame, the study of the reaction of formaldehyde and phenol that led to the formation of "Bakelite." He repeated each step where his predecessors had failed to find the answers. They used low heat; he used high heat. They used low pressure; he used high pressure. They used acid; he used alkali. The result was the granting of the first United States patent for a thermo-setting resin on December 7, 1909. Baekeland did not allow the use of his patent on a royalty basis but instead organized factories in both the United States and in Europe under the banner of the Bakelite Corporation, later to become a unit of Union Carbide and Carbon Corporation. Bakelite was the first of a long series of resins that would shape the economy of the United States, including such resins as cellophane in 1912, acetate in 1927, vinyl in 1928, plexiglas in 1930, acrylic in 1936, melmac in 1937, styrene in 1938, formica in 1938, and polyester and polyamide in 1940.

Baekeland received many honors, among these were the Franklin Medal and being president of the American Chemical Society. He died February 23, 1944, at Beacon, New York.

Barbed Wire

On November 24, 1874, Joseph F. Glidden, a farmer from Illinois, received the first United States patent for a commercially acceptable barbed wire. The patent No. was 157,124.

Many other United States patents previous to Glidden's had been granted for wire fencing. These included patents to S. F. Dexter on August 19, 1833; H. Jenkins, No. 6,106, on February 13, 1849; J. B. Reyman, No. 18,301, on September 29, 1857; and J. W. Rapplege, No. 116,755, on July 4, 1871. Even previous United States patents had been issued for barbed wire fencing to Lucien B. Smith, No. 66,182, on June 25, 1867; William D. Hunt, No. 67,117, on July 23, 1867; Michael Kelly, No. 74,379, on February 11, 1868; and Henry M. Rose, No. 138,763, on May 13, 1873. However, it was Glidden who received the first United States patent for a barbed wire that achieved great commercial success. Within 25 years, the annual tonnage of this new fencing grew to about 250,000.

Joseph Farwell Glidden was born in Charlestown, New Hampshire, on January 18, 1813. Shortly thereafter, his family moved to a farm in Orleans County, New York. Here, he attended school and worked on the farm. Later, he attended college at Middlebury Academy in Vermont and the seminary in Lima, New York. For the next several years, he taught school and worked on his father's farm. In order to acquire the necessary funds to purchase his own farm, he bought two threshing machines and for two years worked his way with these machines into Illinois. In De Kalb County, Illinois, Glidden purchased 600 acres of land and became a prosperous farmer.

He became friends with Isaac Ellwood, a neighboring farmer, and Joseph Haish, a building contractor and lumber dealer. In the summer of 1873, Glidden and his friends attended the county fair in De Kalb and observed an exhibit of barbed wire recently patented by Henry M. Rose. Improvements occurred to both Glidden and Haish. Both men independently filed for a patent in 1873, with Glidden filing two months before Haish. The Patent Office instituted

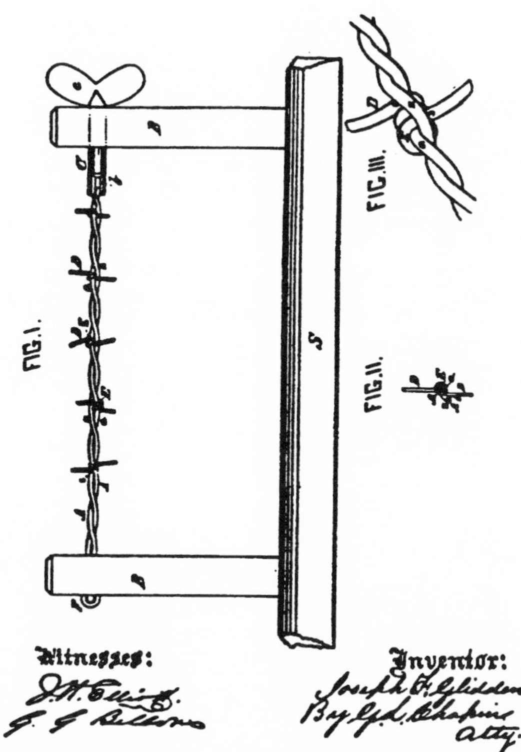

Joseph F. Glidden's design for barbed wire, patented November 24, 1874

interference proceedings with a decision rendered in favor of Glidden on October 20, 1874. Haish then invented the so-called "S" barbed wire, for which he received United States Patent No. 167,240 on August 31, 1875, and proceeded to manufacture it in De Kalb.

Glidden was subsequently granted two other patents on improvements to his barbed wire. He offered to sell a half-interest in his barbed wire to a neighbor for $100 and was refused. However, in 1875, Glidden did sell this same half-interest to his friend, Isaac Ellwood, for $265. They formed a partnership known as the Barb Fence Company, and later manufactured barbed wire in De Kalb, using an old coffee mill converted into a twister to make the barbs. Apparently, both Glidden and Haish sold all the wire they could make.

Meanwhile, Charles Washburn of the Washburn and Moen Manufacturing Company of Worcester, Massachusetts, noticed that large orders for iron wire were coming from De Kalb, Illinois. Being an enterprising industrialist, he went to De Kalb in the spring of 1876 to find out the need for such large orders of wire. After observing what was going on, he offered Haish $200,000 for his patent and factory. Haish refused, wanting more money. Washburn then turned to Glidden and bought his remaining half-interest for $60,000 plus a royalty of a quarter of a cent per pound on wire subsequently sold by Washburn and Moen. Washburn arranged that Ellwood should also make Glidden wire in De Kalb under the condition that Ellwood should sell it only in the West.

The validity of the Glidden patent was upheld by several District Courts in five separate suits for patent infringement, and this holding was affirmed by the Supreme Court in Washburn and Moen Manufacturing Company vs. The Beat 'Em All Barbed Wire Company. This decision was recorded in 143 U.S. 275 (1891).

Glidden continued to draw royalties from his patent until 1891, the year in which his patent expired. His other business interests included his farm in De Kalb, a ranch in Texas, the De Kalb National Bank, the De Kalb Roller Grist Mill, and the Glidden Hotel. He died October 9, 1906, in De Kalb. However, he lived long enough to see, partly as a result of his invention, the war between the cattlemen and the farmers subside.

Beehive

The first United States patent for a practical and successful movable-frame beehive, which revolutionized not only all hives but also methods for keeping bees, was granted to Lorenzo Lorraine Langstroth, No. 9,300, on October 5, 1852.

The use of beehives can be traced to the Greeks who used them for many years before the practice spread to the rest of Europe. In these early beehives, bars of wood were suspended across the tops of the hives that permitted removal of individual combs. In 1765, John Geddle of Scotland placed a small straw ring under the hive that allowed for separate spaces of the brood and the honey. Toward the end of the 18th century, Francois Huber, a blind Swiss naturalist, devised hinged frames that spread apart like the pages of a book so the interior of the bee colony could be observed. Early in the 19th century, Johann Dzierzon, a Polish-German clergyman and beekeeper, devised an opening at the rear of the hive in which wooden bars were suspended where the bees could build the comb. Baron von Berlepsch of Germany improved the work of Dzierzon by constructing frames for the combs.

Beekeeping arrived in America with the early settlers. In 1640, the town of Newbury, Massachusetts, established a municipal apiary, and by the end of the 18th century, bee colonies were common. Early apiarists who received United States patents were T. Stanley on December 9, 1811; E. Blake on

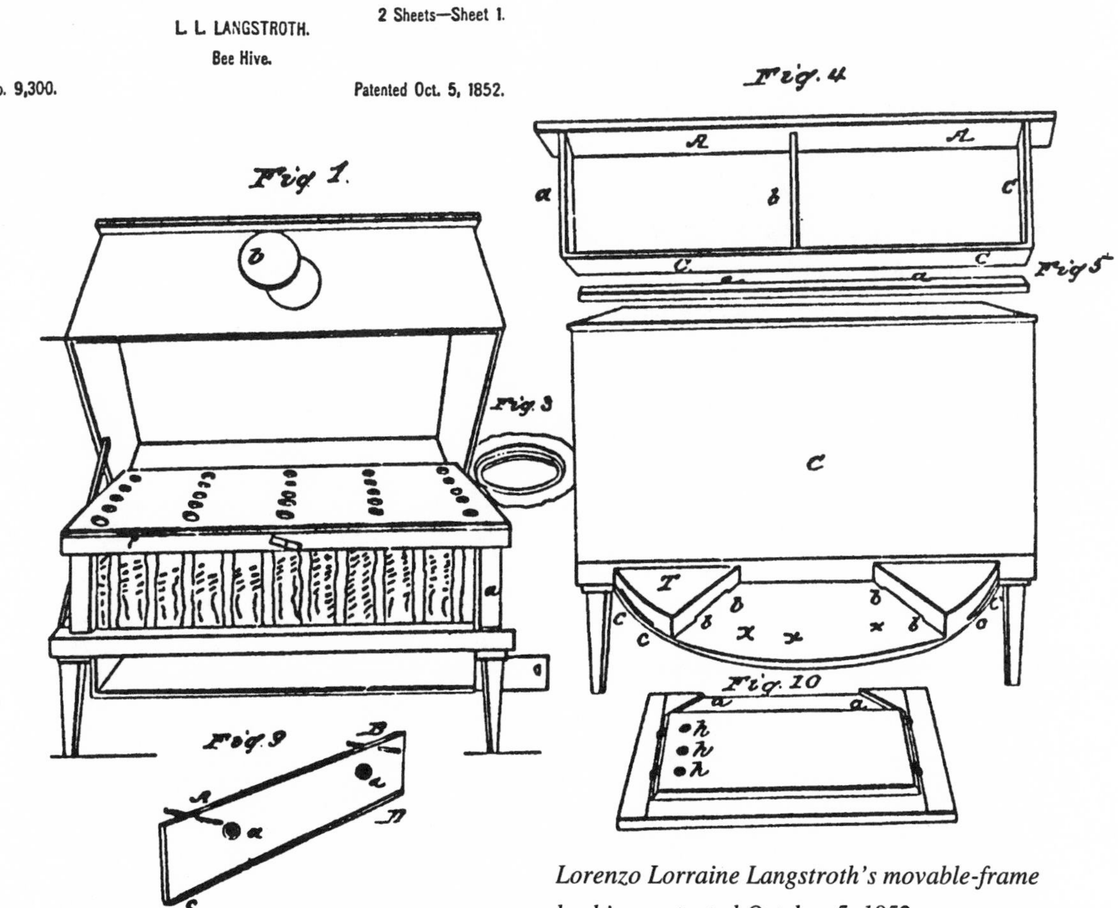

Lorenzo Lorraine Langstroth's movable-frame beehive, patented October 5, 1852

November 16, 1820; S. C. Williams on May 14, 1823; and S. Morrill on February 4, 1834. Two apiarists who deserve particular mention are Moses Quinby of New York and Ambrose Cutting of Haverhill, New Hampshire. Quinby published much material on beekeeping and Cutting received United States Patent No. 3,638 for improving beehives on June 24, 1844. However, they experienced only partial success with their beehives. It was Langstroth whose invention of the bee space formed the basis of all modern bee colony management.

Lorenzo Lorraine Langstroth was born December 25, 1810, in Philadelphia, Pennsylvania. Little is known of his early years except that he graduated from Yale College in 1831. He tutored in mathematics at Yale from 1834 to 1836. During this period, he took a course in theology and became pastor of the South Congregational Church in Andover, Massachusetts, in May of 1836. It was there that he secured two hives of bees and became interested in beekeeping. He remained at South Congregational Church until 1838, when he was compelled to resign because of ill health. For the next five years, he was principal of the Abbot Academy in Andover and then principal of the Greenfield Massachusetts High School for Young Ladies. In 1844, he resumed his pastoral work at the Second Congregational Church in Greenfield. He resigned four years later to become principal of a school for young ladies in Philadelphia. In 1852, he moved to Oxford, Ohio, and took up the work of beekeeping for which he is best known.

Langstroth's invention of the movable-frame beehive was based on his discovery of the bee space. He observed that if a small space, approximately one-fourth of an inch, was left open between the hive and the frame and between the frames themselves, the bees would not fill these spaces with bee glue or combs. Thus, the frames are always movable.

Besides inventing a hive and movable frame, Langstroth was a pioneer in many methods of bee management. These were revealed in his book, *Langstroth On the Hive and the Honeybee*, first published in 1853. With this publication, the United States Patent Office was deluged with applications for new hives. Every imaginable variation of the inventor's design was attempted, many to avoid infringement on Langstroth's patent. Although he knew his patent was clearly valid, Langstroth had neither the financial resources nor the physical strength for the long court battles that followed. So great was the persecution of being robbed of the fruits of his invention that the inventor suffered severe mental distress, refusing for months at a time to see friends, or to talk on the subject of bees. Fortunately, he recovered and lived to see the day when his invention received almost universal adoption. On October 6, 1895, Langstroth died at the home of one of his daughters in Dayton, Ohio.

Bicycle

The first United States patent for a pedal driven bicycle, No. 59,915, was issued to Pierre Lallement on November 20, 1866. The patent was later reissued as Reissue No. 7,972 on November 27, 1877.

The first known depiction of a bicycle appeared in a 1493 drawing by Leonardo da Vinci. Three-hundred years later in 1791, the bicycle was introduced in Paris by Chevalier de Sivrac and was called the dandy-horse. Joseph N. Niepce, the French pioneer in photography, improved the Sivrac machine in 1816 by enlarging the wheels. Also in 1816, Baron Karl von Drais, a German state forester, introduced the Draisine, a bicycle with a pivoted front wheel that could be turned by a handle. It was propelled by thrusting the feet alternately against the ground. It was patented in France for the Baron by Louis Joseph Dineur. In 1818, Denis Johnson patented a similar bicycle in England called the pedestrian curricle.

Bicycles were introduced into the United States in 1819 and were manufactured by David and Rogers in Troy, New York. In the same year, William K. Clarkson of New York received the first United States patent for a velocipede. Little is known about this machine because the patent record was destroyed in the Patent Office fire of 1836 and never restored.

In 1821, Louis Gompertz patented a bicycle in England in which the front handle could propel the machine by a rachet gear. Later, in 1839, Kirkpatrick Macmillan, a Scottish blacksmith, invented the first bicycle to be propelled by mechanical action, and it is considered to be the forerunner to the modern bicycle. It was driven by treadles that moved up and down by pushing with the feet and were connected with long levers to impart a rotary motion to the rear wheel. There is some historical confusion about whether in 1861 it was Ernst Michaux, a coachbuilder in Paris, or Pierre Lallement, an employee of Michaux, who first added pedals to the front wheel of a bicycle. However, it is a fact that the first United States patent for a pedal driven bicycle was awarded to Pierre Lallement on November 20, 1866.

The fact that the Lallement patent is entitled Velocipede is not surprising, because bicycles were then known as velocipedes, curricles, or swift-walkers. The word "bicycle" first appears in an English patent of J. J. Stassen, dated April 8, 1869.

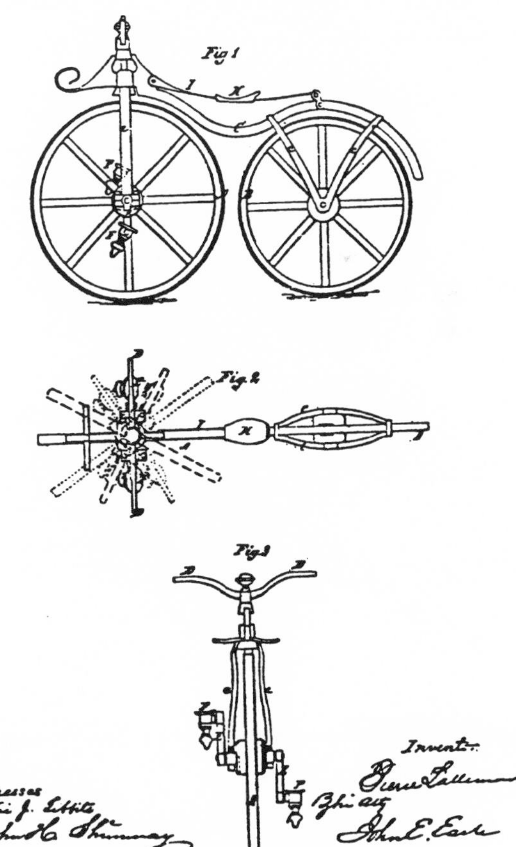

The velocipede, a pedal driven bicycle, patented by Pierre Lallement on November 20, 1866

Pierre Lallement was born in Port a Mousson, France. Little is known of his early years except that in 1863 he went to Paris and found employment with M. Michaux, a carriage builder. It was here that Lallement became interested in improving the bicycle by adding cranks and pedals to an old bicycle owned by his employer. It is not known whether the idea was that of Lallement or Michaux, but Lallement made no effort to secure a patent. The idea surfaced again when Michaux exhibited the machine at the 1865 Paris Exhibition. Shortly thereafter, Lallement departed for the United States, working for his passage as a stoker on a freighter. He arrived in Ansonia, Connecticut, in 1866 and began looking for a job. While searching for employment, he made one of his bicycles and created some excitement by riding it on the streets of New Haven.

An observer named James Carroll recognized that the Lallement bicycle was something novel and likely to be valuable and induced the young Frenchman to patent the invention. A patent was granted on November 20, 1866, to Lallement, who assigned it to himself and James Carroll. This was the first United States patent for a mechanically propelled bicycle. Neither Lallement nor Carroll had sufficient funds for producing the bicycle, so the patent was acquired by Calvin Witty of New York, who began production.

Lallement returned to France and started manufacturing bicycles and continued to do so until the Franco-Prussian War forced a halt to all bicycle manufacturing in France. Later, Lallement returned to the United States and worked for many years for the Pope Manufacturing Company in Boston, makers of bicycles. In 1896, Lallement died, almost unknown and practically forgotten.

Billiard Ball

The survival of the elephant is perhaps due to John Wesley Hyatt Jr. more than any other single individual. For it was Hyatt who discovered a substitute for ivory that could be used in the manufacture of billiard balls and other commercial articles made from ivory. For his discovery, Hyatt was awarded the first United States patent for a billiard ball. The patent was No. 50,359, issued October 10, 1865.

The modern plastics industry probably started with Henry Braccanot, professor of natural philosophy at the Lyceum in Nancy, France. In 1832, he was the first to take cellulose nitrate and chemically form a hard plastic called xyloidine. In 1838, Theophile Jules Pelouze of the University of Paris continued the work of modifying cellulose nitrate to produce a plastic called pyroxylin. In 1846, C. F. Schonbein, a German professor at the University of Basel in Switzerland, discovered the process of nitrating cellulose to make guncotton. In 1847, Waldo Maynard of Boston, Massachusetts, dissolved guncotton in ethyl alcohol and ether to produce a substance called collodion. In 1865, Alexander Parkes of Birmingham, England, was the first to make imitation ivory by combining collodion and camphor. However, it was John Wesley Hyatt Jr. who combined paper flock, shellac, and collodion under optimum heat and pressure to produce a plastic that could be molded into a variety of commercial products, including billiard balls.

No previous United States patent had been issued for a billiard ball. Hyatt's work in this area included not only the first United States patent but also other later patents that issued on April 14, 1868, as No. 76,765, on April 6, 1869, as No. 88,634, and on May 16, 1871, as No. 114,945.

John Wesley Hyatt Jr. was born November 28, 1837, at Stalkey, New York. His father was a blacksmith in the area. After attending common school, he spent a year at Eddytown Seminary studying mathematics. He went to Illinois at the age of 16 and became a journeyman printer. It was there that he received his first United States patent, No. 31,461, for a device for sharpening knives on February 19, 1861. In 1863, he took a job as a printer in Albany, New York, and saw in the newspaper an offer for a prize of $10,000 by Phelan and Collander of New York to anyone who might find a replacement for

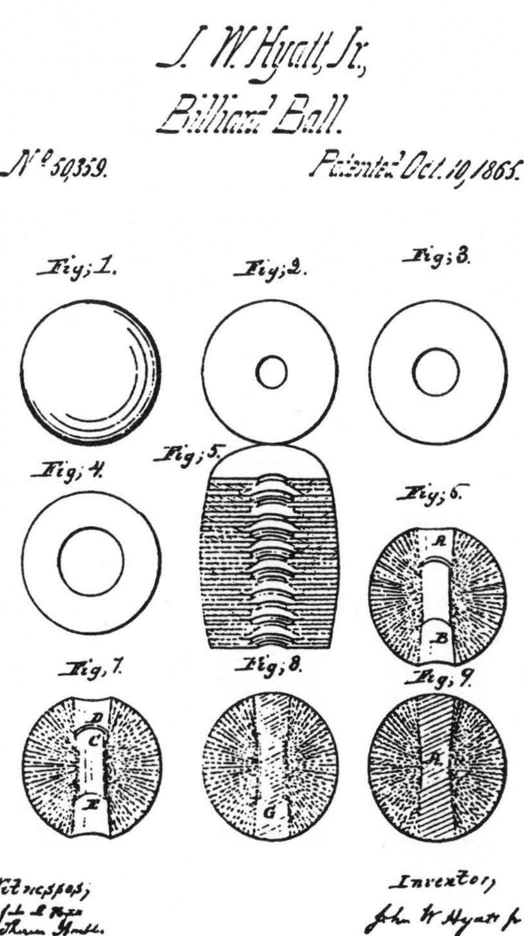

While working on a substitute for ivory in billiard balls, John Wesley Hyatt Jr. developed substances for manufacturing checkers and dominos.

ivory in making billiard balls. It was only right for this prize to be offered by Phelan and Collander, because Michael Phelan had won the first national billiards match at Detroit, Michigan, on April 12, 1859 and became known as the "father of billiards." He therefore had a keen interest in the continuation of the game.

Hyatt set out to win that prize. He converted a shed in the rear of Mrs. MacTavish's boardinghouse into a workshop and with fellow printer, James Brown, built a machine in which, by applying heat and pressure, they hoped to mold billiard balls out of pulverized wood, paper pulp, and shellac. The process worked poorly for the manufacture of billiard balls but was successful for manufacturing checkers and dominoes. In fact, Hyatt and his two brothers established the Embossing Company of Albany for the manufacture of these materials.

Hyatt continued to seek a substitute for ivory in the making of billiard balls. Noticing that an overturned bottle of collodion had formed an ivory looking hard blob on the shelf, he achieved success on finding a substance suitable for billiard balls using paper flock and shellac to form a ball under heat and pressure and then coating the surface with collodion. Although billiard balls were produced commercially by this method, they had certain disadvantages. A lighted cigar applied to the ball sometimes resulted in a serious flame and occasionally the balls on striking one another would produce a mild explosion.

Hyatt, together with his brother Isaiah, continued experimenting with materials subjected to high temperature and pressure. One of these materials was a mixture of pyroxlin and camphor which yielded a superior product, not only for billiard balls but for hundreds of other products. Isaiah named this product Celluloid and together they were awarded United States Patent No. 105,338 on July 12, 1870. Daniel Spill, the English inventor of xylonite, brought suit against Hyatt for the prior rights to celluloid stating he was the first to make this product. However, the courts thought differently and awarded priority to Hyatt. In early 1873, the Hyatts moved to Newark, New Jersey, and started the Newark Celluloid Manufacturing Company. They also obtained United States Trademark No. 1,102 on the word Celluloid on January 14, 1873.

Hyatt's inventive versatility resulted in his being granted over 200 patents for a variety of inventions. Some of these were extremely important and resulted in establishing the Hyatt Pure Water Company and the Hyatt Roller Bearing Company at Harrison, New Jersey. He also invented a sugar-cane mill, a sewing machine capable of sewing 50 lockstitches at once, a machine for cold rolling and straightening steel shafting, a machine for making a slate blackboard for use in schools, and a machine for turning out billiard balls. The Society of Chemical Industry of London in 1914 awarded Hyatt its Perkin Medal, a distinguished honor, particularly since he was never a chemist.

There is no evidence that the $10,000 prize was ever paid to Hyatt or anyone else. However, the inventor was more than compensated for this oversight when he became known as the "founder of the plastics industry in the United States." He died at Short Hills, New Jersey, on May 10, 1920.

Bottle Cap

The search for a better bottle closure led William Painter to invent a bottle cap that represented a major milestone in the carbonated beverage industry. He received the first United States patent, No. 468,226, on February 2, 1892, for a bottle cap with the crown seal that was adopted worldwide.

The search for a bottle closure had been in progress for many years before the Painter patent. H. N. DeGraw was awarded United States Patent No. 14,446 on March 18, 1856, for a ready-made cork holder. Henry W. Putnam received United States Patent No. 23,263 on March 15, 1859, for a wire clamp to hold the cork in the bottle. Albert Albertson was awarded United States Patent No. 36,266 on August 16, 1862, for a metal spring inside the bottle to activate the closure. John Matthews was issued United States Patent No. 48,422 on June 25, 1865, for an internal stopper, an elongated plug which closed from the pull of a magnet. Hiram Codd received United States Patent No. 138,230 on April 29, 1873, for a device in which a glass ball was held against a rubber gasket by internal pressure. Charles G. Hutchinson was granted United States Patent No. 213,992 on April 8, 1879, for a spring-type bottle stopper. William L. Roorback was awarded United States Patent No. 272,775 on February 20, 1883, for a floating-ball stopper. Alfred L. Bernardin was issued United States Patent No. 314,358 on March 24, 1885, for a metallic cap and fastener. The many patents on bottle closures, all of which had limited commercial success, was evidence that a better bottle closure means was needed. This need was met by William Painter and his bottle cap, a metallic disk with a cork liner that is securely crimped on the bottle top.

William Painter was born at Triadelphia, Montgomery County, Maryland, on November 20, 1838. During the first 10 years of his life, his father farmed in various places in Maryland, the last being at Fallston in Harford County. His early education was received in Friend's schools in Fallston and in Wilmington, Delaware. In 1855, he became an apprentice at Pyle, Wilson & Pyle, a patent-leather manufacturing plant in Wilmington, and remained there for four years. While there, he invented and patented a fare box on August 3, 1858, and a railroad car seat on August 31, 1858. In 1859, he returned to Fallston, Maryland, and worked as his father's assistant in a general store. During this time, he patented

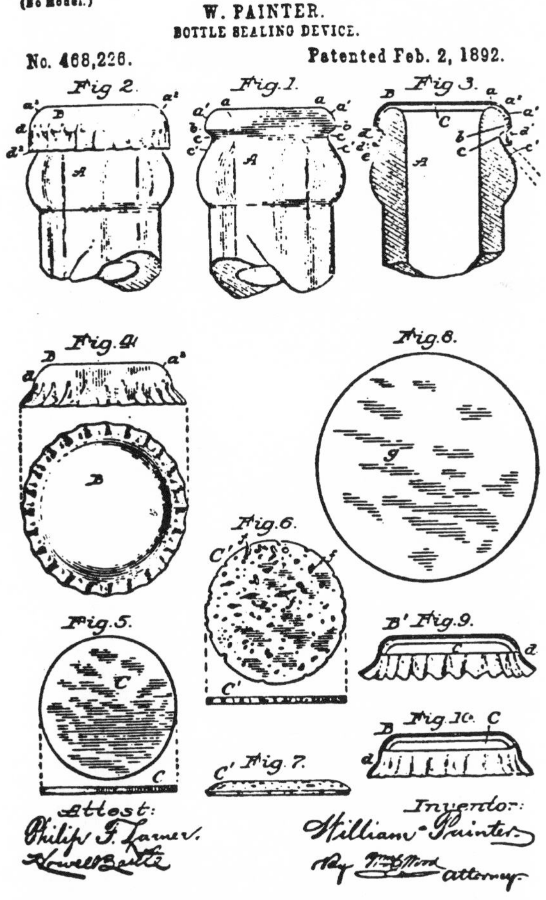

A now common sight, and a great idea—a bottle sealing device by William Painter, patented February 2, 1892

a counterfeit-coin detector on July 8, 1862, and a kerosene lamp burner on June 30, 1863. In 1865, he moved with his family to Baltimore and accepted the position of foreman of the Murrill & Keizer machine shop. He remained there for the next 20 years, inventing more than 35 devices, including an automatic magneto-signal for telephones, a seed sower, a soldering tool, and several pump valves.

In 1880, Painter turned his attention to bottle stoppers and on April 14, 1885, received United States Patent No. 315,655 for a wire-retaining rubber stopper. This became known as the Triumph swing-type bottle stopper. To market this invention, the Triumph Bottle Stopper Company was organized by Painter and some friends. On September 29, 1885, he obtained United States Patent No. 327,099 for a single-use closure bottle seal. Since this bottle seal could be made and sold for 10 times less than the Triumph stopper, the Bottle Seal Company was organized to replace the Triumph Bottle Stopper Company. Painter then received United States Patent No. 468,226 on February 2, 1892, for a single-use cap of metal with a cork liner. This became known as the Crown Cork bottle cap and is used extensively throughout the world today. To market this latest invention, the Bottle Seal Company was reorganized as the Crown Cork and Seal Company on March 9, 1892. Painter later obtained patents on improvements to the crown cap and for a crown opener.

During his career, Painter was granted 85 patents, the last one being issued after his death. He died in Baltimore on July 15, 1906.

Bridge Caisson

The bridge crossing the Mississippi River at St. Louis, Missouri, is appropriately named the Eads Bridge. It was built under the supervision of James Buchanan Eads who introduced pneumatic caissons in bridge building in the United States. He was awarded the first United States patent, No. 123,002, for a pneumatic caisson on January 23, 1872.

Pneumatic caissons were first suggested in 1647 by French physicist Denis Papin; in 1779 by French scientist Charles Augustin De Coulomb; and in 1841 by French engineer M. Triger. They were first used in bridge construction by Lewis Cubitt and John Wright at Rochester, England, in 1851. Other bridges erected on pneumatic caissons were the Royal Albert Bridge by Sir Marc Isambard Brunel at Saltash, England, in 1855; the Rhine Bridge at Kehl, Germany, in 1859; and a bridge at Vichy, France, in 1869. It was James Buchanan Eads who perfected and introduced the pneumatic bridge caisson in the United States.

James Buchanan Eads was born May 23, 1820, at Lawrenceburg, Indiana. His father, Thomas, was a merchant. The family moved to Cincinnati, later to Louisville, and finally to St. Louis in 1833 when Eads was 13 years of age. His schooling ceased at this time and he peddled apples on the street to help support the family since his father's mercantile business was not prospering. He then found employment as a clerk in the dry-goods store of Williams & Durings and spent most of his spare time in the Williams library. In 1838, Eads secured a job as a cargo clerk on the Mississippi River steamboat Knickerbocker for three years. While there, he invented a diving bell with which he spent many hours studying the bottom of the Mississippi. In 1842, he formed a partnership in steamboat salvaging. The business became successful, and three years later, he sold his interest and built a glass factory in St. Louis, the first glass factory west of the Ohio River. This business, however, was a failure, and in 1848, Eads found himself in debt. He then returned to the salvage business of raising sunken steamers, and by 1857 had not only paid off his $25,000 debt but also had

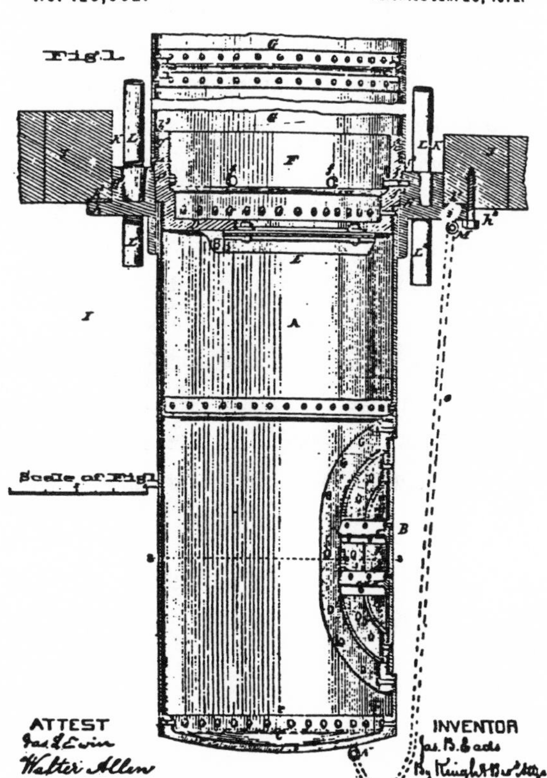

An improvement in the construction of subaqueous foundations, patented by James B. Eads on January 23, 1872

amassed a fortune that allowed him to live in semiretirement from 1857 to 1861.

During the Civil War, Eads, at the invitation of President Lincoln, built a fleet of 14 armor-plated steam-propelled gunboats for the government. In

1865, a bill was introduced in Congress authorizing the construction of a bridge across the Mississippi River at St. Louis. Eads was selected engineer-in-chief with Colonel Henry Flad as his first assistant. Actual construction began on August 20, 1867. Since the center arch was to be 520 feet long and the two other arches were to be 502 feet each, and since the base of one pier was to be 136 feet below high water, and was to be sunk through 90 feet of sand and gravel to bedrock, a pneumatic caisson was a necessity. Eads built a cylindrical iron-shod caisson of heavy timbers, reinforced with iron bands. The diameter was 75 feet and the working chamber 8 feet deep. Within this airtight enclosure, submerged in water, men could work unimpeded by tides of shifting sands. Sand pumps, also invented by Eads, lifted and ejected gravel, sand, and silt from the caisson chamber. Also within this enclosure or caisson the masonry pier was built up, the weight of the masonry forcing the cutting edge into the river bed. The caisson extended continuously up to the water surface where successive rings were added as it sank lower.

Five months of excavation and pumping were required to uncover the foundation rock for the east pier. With the completion of the three piers in 1873, the construction of the superstructure required only one-fifth of the time of the piers. Not only was the use of a pneumatic caisson a first, but also was the first extensive use of steel in bridge construction and the first use of hollow tubular chord members. The bridge, which contained a roadway on the upper deck and a railroad line on the lower deck, was completed and opened to traffic in 1874 at a cost of $6.5 million.

After completing this bridge, Eads began work to open one of the mouths of the Mississippi River, the South Pass, into the Gulf of Mexico. There, he had the opportunity to apply his knowledge of the river's currents. He dug a system of jetties that allowed the river to deposit sediments thereby causing the remaining water to dig a deeper channel so large ships might come up the Mississippi to St. Louis. He successfully accomplished this task in 1879.

Many honors came to Eads. In 1877, the University of Missouri conferred upon him an honorary doctorate, and in 1884, the British Society for the Encouragement of Art, Manufacture and Commerce awarded him the Albert Medal. Eads was the only American up to that time to receive this award. In 1920, he was elected to the American Hall of Fame, the first engineer to be thus honored. In January of 1887, he went to Nassau, Bahamas Islands, for his health and died there March 8, 1887.

Cable Car

At one time in the history of the United States, the cable car was the most economical available solution to the urban transportation problem. Andrew Smith Hallidie, more than any other individual, was responsible for this solution. Drawing on a combination of existing technological devices, he invented a commercially successful cable car and was awarded United States Patent No. 110,971 on January 17, 1871.

Vehicle movement by cable traction was not new in 1871. Many men previous to Hallidie had experimented with cable traction. Early work on cable traction movement was performed by W. and E. K. Chapman, who proposed moving vehicles along streets by a fixed cable in 1812. William James, in 1824, used a chain in a hollow rail to move vehicles along highways. M. Dick experimented with an endless-cable system with a stationary power source in 1829. W. H. Curtis studied a rope haulage system in which the vehicle was clamped and unclamped by vertically moving jaws in 1838. E. W. Brandling in 1845 proposed the first system based on a cable moving within an open box. On March 23, 1858, E. S. Gardner received United States Patent No. 19,736 on a conduit with a narrow slot and carrying pulleys for guiding a cable. Charles T. Harvey in 1866 patented a system based on gripping forks or claws on rapid-transit cars to engage an open-running steel cable. The former Confederate General George Beauregard in 1870 experimented with overhead cable for his New Orleans & Carrollton Railroad. However, it was Hallidie who combined these ideas to invent a successful cable car moved by cable traction in 1871.

Andrew Smith Hallidie was born in London on March 16, 1836, as Andrew Smith. By family consent, he adopted the surname Hallidie in honor of his godfather and uncle, Sir Andrew Hallidie, a prominent Scottish physician. His father, Andrew Smith, was an inventor who had received a British patent on wire rope in 1835. After some education and apprentice training in civil engineering and surveying, young Hallidie went to California in 1853 and became a prospector in the Sierra. Unsuccessful in seeking gold, he was engaged to build a flume across

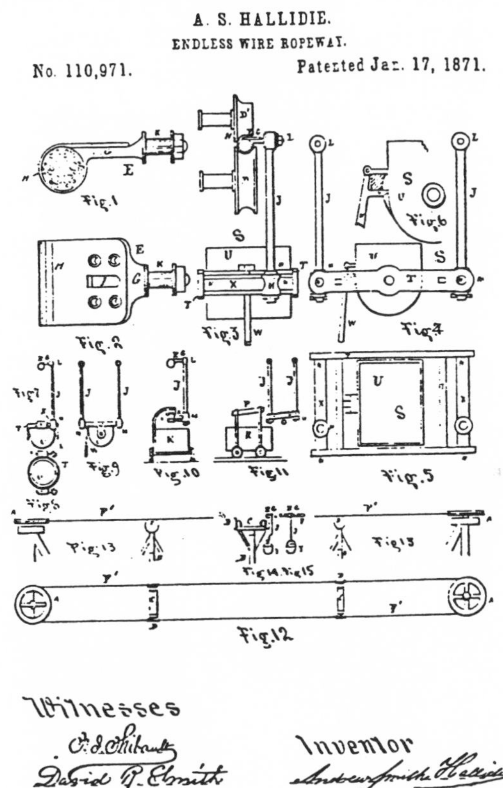

Andrew Smith Hallidie's cable car, patented January 17, 1871

the middle fork of the American River. There, he designed and built a wire suspension structure to carry an open flume for a mine at American Bar. This structure was 220 feet long and was very successful. In 1860, he formed the A. S. Hallidie Company in San Francisco and for the next 12 years he designed and built at least 14 wire suspension bridges and flumes. In 1857, he established a small factory for wire rope in San Francisco that subsequently developed into the California Wire Works. For the next 10 years, he divided his energies between the cable factory and the designing of bridges and haulage systems for mines. In 1867, he patented the Hallidie Ropeway, a system for transporting freight and ore in and out of mines by endless wire ropes.

The success of his ropeway suggested to Hallidie the application of the same principle to the pulling of loaded streetcars up the steep hillsides of Nob Hill in San Francisco. By 1871 he had devised and patented an underground endless moving cable and a mechanical gripping device to be attached to the underside of the streetcar. With the assistance of Joseph Britton, Henry L. Davis, and James Davis, the Clay Street Hill Railroad opened for business on September 1, 1873. The fares were five cents and earned for Hallidie about $3,000 a month. In addition, Leland Stanford paid Hallidie $30,000 for the use of his patents in the construction of other San Francisco lines.

The cable car industry soon spread to other American cities. Other engineers who worked on street railways at the same time as Hallidie or immediately following were William Eppelsheimer, Henry Casebolt, W. H. Paine, Asa Hovey, Leland Stanford, Henry Root, and C. B. Holmes. Nearly 4 million passengers used cable car service annually in 1893 in 28 large cities of the United States.

The 1906 earthquake in San Francisco and the invention of electric street trolleys and diesel-powered buses caused the demise of the cable car in San Francisco and other cities. The final run of the original Clay Street cable car was on February 15, 1942. Other San Francisco cable car service continued until 1982 when all cable car service was stopped. However, due to increased public pressure, the city completed a $60 million restoration project in 1984 which resulted in three cable car lines being restored and is today one of the city's main tourist attractions. The cable cars of San Francisco are also registered as National Historic Landmarks.

Hallidie, in addition to his wire rope and cable car interests, was quite active in civic affairs. He was a regent of the University of California from its founding in 1868, president of the Mechanics Institute of San Francisco, vice president of the James Lick School of Mechanical Arts, and a founder of the San Francisco Public Library and Art Society. He died in San Francisco on April 24, 1900.

Calculator

William Seward Burroughs made the task of bankers, bookkeepers, and cashiers much easier. He was the first to receive a United States patent for a machine that would not only calculate but also record and print the result. For this, he received United States Patent No. 388,116 on August 21, 1888.

The first calculators were human fingers. Next came the abacus in China and the soroban in Japan. Then in 1614, John Napier, a Scottish mathematician, arranged strips of bones on which were figures so they could be brought into various fixed combinations. These were called Napier's bones and eventually evolved into the slide rule. Blaise Pascal, a French mathematician, made the first arithmetical machine that embodied wheelwork in 1641. Sir Samuel Morland, an English mathematician, invented a stylus-operated machine with no automatic carry from units to tens in 1661. Gottfried Liebnitz, a German mathematician, invented the first machine to perform multiplication by means of successive additions in 1694. Charles Xavier Thomas de Colmar, a native of Alsace-Lorraine, was the first to make a satisfactory commercial calculating machine in 1820 called the Thomas de Colmar Arithmometer. Charles Babbage, an English mathematician, started to build an automatic calculating machine in 1833 called the difference engine or analytical engine. It was only partially completed when he died in 1871. George and Edward Scheutz, of Stockholm, Sweden, built a smaller version of the Babbage machine in 1853.

The first United States patent for a calculating machine was issued to O. L. Castle of Alton, Illinois, in 1850. It was for a 10-key adding machine that did not print and only added in one column. In 1875, a United States patent was granted to Frank S. Baldwin for a calculating machine that was not very successful in this country but for which he was awarded in 1875 the John Scott Medal by the Franklin Institute. It remained for William S. Burroughs to perfect the calculator-recorder that eventually received public acceptance.

William Seward Burroughs was born January 28, 1855, in Auburn, Cayuga County, New York. His father, Edmund, was a model-maker for new inventions. After a limited education in the Auburn schools, young William, at the age of 15, began earning his own living. He found work as a bookkeeper in a small bank in Auburn where he discovered that nine-tenths

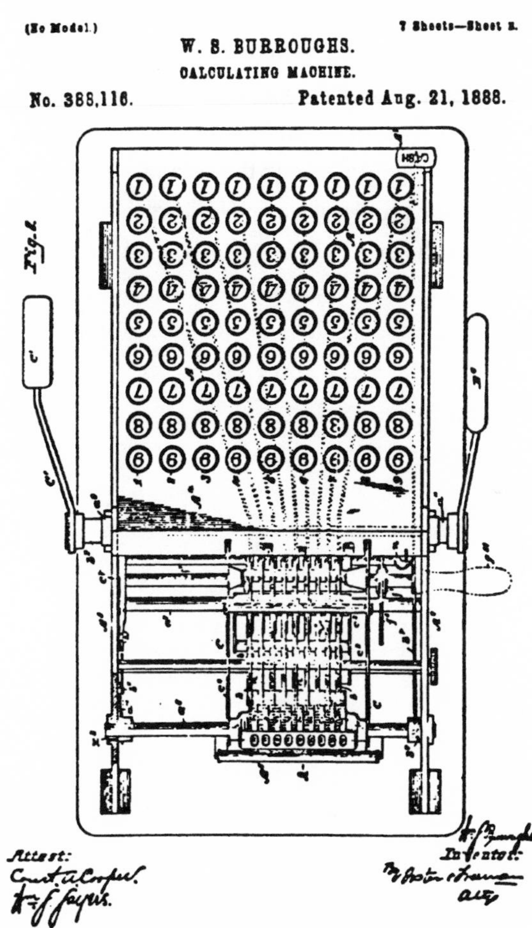

William Seward Burroughs created this machine to calculate, record, and print numerical data

of his work was pencil and paper addition. Long hours and poor health drove Burroughs to start thinking about a machine that would do this work for him. Due to the tedious work, his health gave way under the strain, forcing him to resign. He moved with his family to St. Louis and there for a short time worked in his father's model shop. It was there that he met Thomas B. Metcalf who encouraged Burroughs financially to work on a machine for solving the arithmetical problems like those encountered at the bank. He began working on his machine in the Joseph Boyer Machine Shop.

By the end of 1885, he had completed a machine. These machines were made by the American Arithmometer Company in St. Louis that had been organized by Burroughs, Metcalf, and two St. Louis merchants. Unfortunately, they were not successful at first. The handle, when pulled down too suddenly, affected the results. It had to be pulled forward slowly and then released. The public, not knowing how to use the handle, lost confidence in the machine. The company was headed for bankruptcy until Burroughs invented a device in 1890 that eliminated the handle problem and made the machine foolproof. This device was a small cylinder that was partially filled with oil and contained a plunger. This plunger was connected to the handle and acted as a shock absorber. The machine would operate only at a certain speed no matter how the handle was pulled. These new machines, after further improvements, were commercially successful. In 1895, 284 machines were sold, and in 1904, 1,000 machines were in use. In 1905, the Burroughs Adding Machine Company, successor to the Arithmometer Company, was moved to Detroit with Joseph Boyer as president of the new company. By 1913, the company had annual sales of over $8 million, the equal of all its competitors.

Burroughs was awarded the John Scott Medal by the Franklin Institute in 1897. He died September 14, 1898, at his country home in Citronelle, Alabama, a victim of tuberculosis.

Cannon

The cannon has played an important role in the history of many countries. One of the individuals responsible for the development of the cannon was Thomas Jackson Rodman who received the first United States patent for casting guns with a hollow core and water cooling the inner wall of the cast-iron cannon. Such a procedure revolutionized the cannon-making industry. The United States Patent No. was 5,236, issued August 14, 1847.

Cannons were first used about 1250 after the invention of gunpowder. They were featured in many European battles. Edward III of England used them against the French in the Hundred Years War beginning in 1337. The Ottoman Turks under Mohammed II used 68 cannons to capture Constantinople in 1453. The largest had a 30-inch bore and hurled a 1600-pound stone a distance of one mile. Charles VIII of France used the first mobile cannon against Italy in 1495. Niccolo Tartaglia, an Italian mathematician, published a book in 1537 on gunnery and invented the gunner's quadrant, which took part of the guesswork out of aiming.

The cannon was introduced very early in American history. Artillery units made important contributions to American victories in the Revolutionary War and the War of 1812. Ordnance workers before Rodman who had received United States patents for the cannon were J. W. Cochram on March 23, 1836; George Chapman on June 16, 1836; Cyrus Alger on May 20, 1837; Ransom Cook on February 1, 1842; and A. Eames on February 5, 1847. Ordnance workers during or shortly after Rodman who helped in cannon development were Benjamin Chambers who received United States Patent No. 6,612 on July 31, 1849, for a breech-loading mechanism using screw threads and Alfred Krupp of Germany who received United States Patent No. 13,851 on November 27, 1855, for the use of steel in cannon manufacturing.

Other workers who deserve special mention were George Bomford, Robert Parker Parrott, and Lieutenant John A. B. Dahlgren of the United States, and Henri Joseph Paixhans of France. However, it was Rodman's casting method that was accepted not only by the United States but also by many foreign countries, and cannons produced by this process provided for the costal defense in the United States until about 1885.

Thomas Jackson Rodman was born on a farm near Salem, Indiana, on July 30, 1815. After receiving the standard early education, he entered the United States Military Academy in 1837. He gradu-

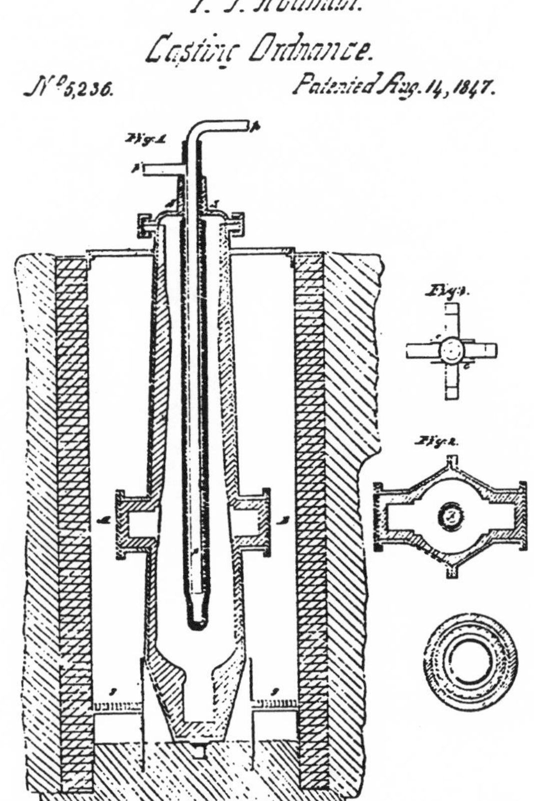

The patent that revolutionized the cannon-making industry, granted to Thomas Jackson Rodman, August 14, 1847

ated in 1841 as seventh in a class of 52 members. He was commissioned in the ordnance department and was promoted to the rank of captain in 1855. His early service was spent with the government arsenals at Pittsburgh, Pennsylvania; Richmond, Virginia; and Baton Rouge, Louisiana. During the Mexican War, he served as ordnance officer at Camargo and Point Isabel.

Rodman studied the heavy ordnance of earlier workers, including George Bomford, Henri Joseph Paixhans, and Lieutenant John Dahlgren. Taking the Dahlgren bottle shape, he improved the method of casting by casting the gun upon a hollow core and cooling it by a current of water made to flow into and out of the bore so each successive layer of metal was compressed by the shrinkage of the outer layers, thereby gaining density of the metal on the interior. The result was a marked increase in strength and endurance for heavy ordnance. The army ordnance department at first rejected his invention, so private enterprise first manufactured the guns under Rodman's patented process. However, the government in 1859 finally approved and adopted his invention as well as the governments of Russia, Great Britain, and Prussia. In 1867, he was promoted to the rank of brigadier-general.

Rodman also experimented with gunpowder used in the testing of his large-bore cannon. He developed the Mommoth powder consisting of large grains compressed to high density. He also developed the perforated-cake or prismatic gunpowder. In 1861, he published a book in Boston that described his work on gunpowder called, *Reports of Experiments On the Properties of Metal for Cannon, and the Qualities of Cannon Powder.*

During the Civil War, he was placed in command of the arsenal at Watertown, Massachusetts, and supervised the casting of 12-inch, 15-inch, and 20-inch guns. At the close of the Civil War, he was transferred to the command of Rock Island Arsenal in Illinois where he died at his post after a prolonged illness on June 7, 1871.

Carborundum

Edward Goodrich Acheson invented a superior artificial abrasive still in use today called Carborundum. He was awarded the first United States patent for a silicon carbide abrasive used throughout the world. The patent was No. 492,767 and was issued on February 28, 1893.

Farmers and metal workers had been sharpening tools and grinding metals many years before Acheson and many others received United States patents on grinding stones and wheels. Patents for grindstones went to J. Fitzgerald, No. 4.755, on September 12. 1846; G. G. Griswold, No. 45,243, on November 29, 1864; and W. Tanner, J. S. Hyatt and J. W. Hyatt, No. 53,056, on March 6, 1866. Those for grinding wheels include E. H. Danforth, No. 48,160, on June 13, 1865; and A. W. Calder, No. 93,413, on August 10, 1869. Others for emery wheels went to J. D. Alvord, No. 30,189, on October 2, 1860; and J. F. Wood, No. 110,102, on December 13, 1870. However, it was Acheson who achieved industrial history by inventing a superior abrasive made from silica and carbon.

Edward Goodrich Acheson was born in Washington, Pennsylvania, on March 9, 1856. His family lived over a grocery store that was run by Acheson's father. When he was five years old, the family moved to Monticello, Pennsylvania, where his father became manager of a blast furnace. Young Acheson's schooling was confined to seven years at the district school conducted by one of the neighboring farmers, one year at a boarding school in North Sewickley, Beaver County, Pennsylvania, and two years at the academy in Bellfonte, Centre County, Pennsylvania. His main interest while at the academy was in mathematics and surveying.

He was called home in 1872 and became timekeeper at his father's blast furnace. It was there that he received his first United States patent on March 5, 1873, for a rock-boring machine to be used in the coal mines, the first of 69 patents he received.

His father died in 1873 and after a succession of temporary jobs, Acheson went to New York and secured a position as a draftsman with Thomas A. Edison at Menlo Park, New Jersey. He assisted with the lamp exhibit at the Paris Exposition in 1881 and spent the next two years installing electric lighting plants in various countries in Europe. He then re-

E. G. ACHESON.

PRODUCTION OF ARTIFICIAL CRYSTALLINE CARBONACEOUS MATERIALS.

No. 492,767. Patented Feb. 28, 1893.

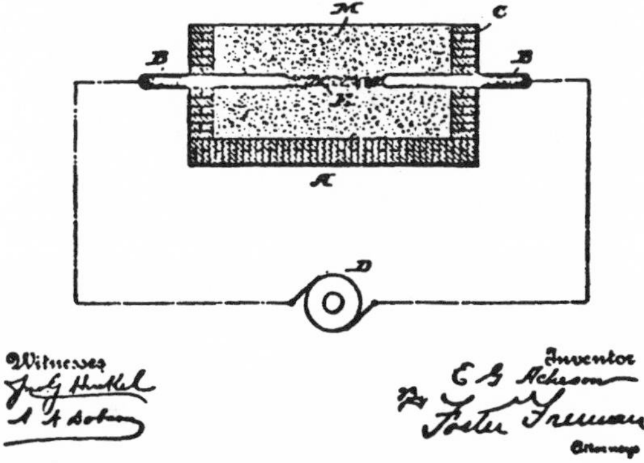

A patent for production of artificial crystalline carbonaoeous materials, granted February 28, 1893

turned to Menlo Park and started his own experimental laboratory. This venture was not successful so he returned to Monticello, Pennsylvania (renamed Gosford), and started experimenting on the reduction of iron ore with unsatisfactory results.

In 1880, while Acheson was working with iron ore, Dr. George F. Kunz of Tiffany & Company of New York, made a remark that a good abrasive was one of the important industrial needs. Acheson decided in March of 1881 to pursue this idea and started to experiment with a mixture of clay and carbon under the influence of electric heat. On examining the cooled melt, he found a few shiny specks that were hard enough to scratch glass. Further experimentation, with the aid of an electric furnace that he developed, led to more of the product. He called the product Carborundum, thinking it was a compound of carbon and corundum, a mineral containing aluminum oxide. Later analysis revealed it to be silicon carbide. Acheson then organized the Carborundum Company on September 21, 1891, which eventually moved to Niagara Falls, New York, in 1895. The company prospered and in 1909 was producing 10 million pounds of Carborundum a year.

Acheson then turned his attention to the manufacture of graphite and organized the Acheson Graphite Company in 1899. This business grew rapidly and produced such articles as electrodes, plates for motor brushes, bulk graphite for dry batteries, and paint pigment. In 1909, the company was producing graphite at the rate of more than 10 million pounds per year. Through an exchange of stock, he transferred in 1928 ownership of the Acheson Graphite Company to the Union Carbide and Carbon Corporation, now Union Carbide Corporation. Acheson was instrumental in successfully establishing at least five industrial corporations in the United States and Europe relating to the electrothermal process.

Acheson received many honors during his latter years. These included the John Scott Medal of the Franklin Institute in 1894 for the discovery of Carborundum and again in 1901 for the production of artificial graphite. He also received the Count Rumford Medal from the American Academy of Arts and Sciences in 1908. The University of Pittsburgh conferred on him the PhD degree in 1909. He received the Perkin Medal in 1910 from the American Section of the Society of Chemical Industry. The King of Sweden appointed him an Officer of the Royal Order of the Polar Star in 1914. In 1928, he gave $25,000 to the American Electrochemical Society for establishing the Acheson Medal to be awarded every two years.

Acheson died of pneumonia at the home of a daughter in New York City on July 6, 1931.

Carpet Loom

The carpet industry gained impetus with the invention of the power loom by Erastus Brigham Bigelow. He received the first United States patent for a power loom to produce carpet. He received United States Patent No. 169 on April 20, 1837.

Many weaving looms were used in the United States before Bigelow's. Some of the more prominent loom inventors were James Davenport in 1764, Samuel Slater in 1790, William Peter Sprague in 1791, Ira Draper in 1808, Francis Cabot Lowell in 1813, Alexander Wright in 1825, Orrin Thompson in 1828, George Crompton in 1836, and Elias S. Higgins in 1840. However, it was Erastus Brigham Bigelow who astonished the world with his superior carpets.

Erastus Brigham Bigelow was born April 2, 1814, in West Boylston, Massachusetts. His father, Ephraim, was a farmer, cotton weaver, and wheelwright. Young Bigelow attended the district school when it was in session and assisted his parents on their farm and wheelwright shop. At the age of eight, he wished to study arithmetic but the teacher of the district school thought that he was not old enough and refused to allow him to enter the class. Between the ages of 10 and 20, he worked on a neighboring farm, played the violin in the orchestra of the Orthodox Society of Lancaster, worked as a clerk in S. F. Morse's retail dry-goods store in Boston, taught penmanship at Newark, wrote a pamphlet called *The Self-Taught Stenographer*, and attended Leicester Academy for a year.

His first successful invention was a power loom for the production of lace in 1837. After the loom for weaving coach lace by power had been invented, the next thing was to put it into practical operation. He and his brother, Horatio N. Bigelow, formed a company on March 8, 1838, called The Clinton Company because of his fondness for the Clinton House in New York. The home of the new plant soon became known as Clintonville, and finally Clinton. While in New York, Bigelow happened to see a new and different kind of counterpane imported from England. He then set about inventing a power loom for

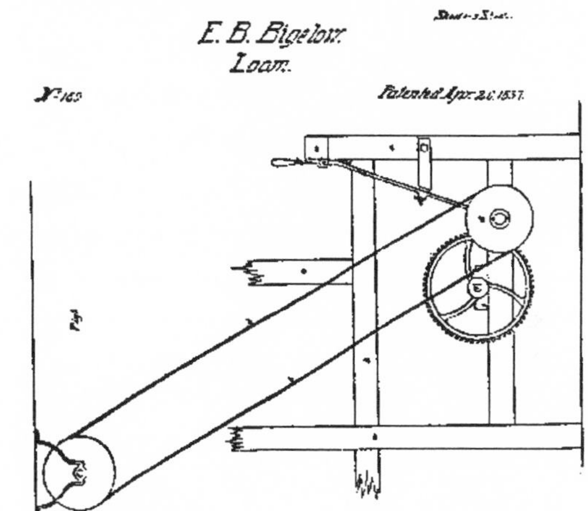

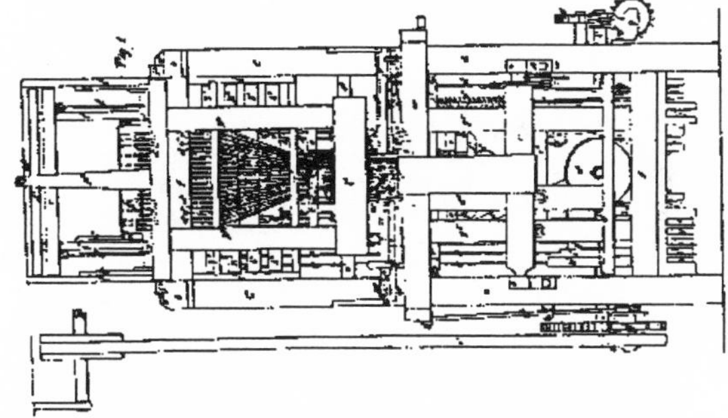

E. B. Bigelow's Carpet Loom, patented April 20, 1837

weaving this new fabric. Within six months, he had the loom completed and operating successfully.

After starting the coach lace and the counterpane establishments, Bigelow took up the problem of weaving Ingrain or Kidderminster carpets by power looms. Before he even made a model or finished a complete set of drawings for the machine, he entered into a contract with Alexander Wright of the Lowell Manufacturing Company to furnish power looms for the manufacture of Ingrain carpets. His first loom for two-ply Ingrains was set up, and in the matching of figures, in the evenness of surface, and in the regularity of selvedge, its products far surpassed that of any hand loom. Its average output was twelve yards a day, while that of the hand loom was eight. A second loom with modifications was made, which raised the daily output to 18 yards. A third loom with certain changes was made with an increase of output to 27 yards per day. The last loom was built in 1841. In 1843, the Lowell Manufacturing Company built a carpet mill at Lancaster covering more than four acres that housed the Bigelow looms, the first successful power-loom carpet mill in existence in the world. *Hunt's Merchant's Magazine* of the period called it "the most perfect establishment in the United States." Bigelow will always be principally remembered as the inventor of power looms for the production of various carpetings.

Bigelow became interested in economics and in 1862 published a book called, *In the Tariff Question Considered in Regard to the Policy of England and the Interests of the United States*. In 1860, he was nominated as a candidate for Congress by the Democrats of the Fourth Massachusetts District but lost by a narrow margin. He was also a leading member of a committee of 21, appointed January 11, 1861, to carry into effect proposals that led to the founding of the Massachusetts Institute of Technology. In 1789, he presented to the Boston Historical Society six volumes of specifications and sketches for patents in his name for various kinds of weavings. Bigelow died in Boston on December 6, 1879.

Clock

Eli Terry received the first United States patent for a clock on November 17, 1797. The patent was awarded for an equation clock, which showed both apparent and mean time.

The sundial is the oldest known device for measuring time. It is believed to have been used in Babylon as early as 2000 B.C. Plato, the Athenian philosopher, is generally credited with the invention of the water clock in 372 B.C. In the early 1300s, Henry de Vick, a German, invented a clock that contained many of the important parts of the modern clock. Jacopo Dondi, an Italian, is credited with designing the first clock dial in 1344. Peter Henlein, a German locksmith, invented the mainspring in 1470, which made possible a portable clock and later the wrist watch. Christiaan Huygens, a Dutch scientist, made the first pendulum clock.

The original Eli Terry Clock patent was destroyed in the patent office fire of 1836 and was never restored. A sample of the Eli Terry Clock may be found on page 25 of the April 1989 issue of *American Heritage.*

The Terry name was famous in American clock history. Eli and his sons, Eli Jr., Henry, and Silas contributed to this history, as well as his brother, Samuel, and his sons, Theodore, R. E. and J. B.

Colonial clockmakers before Eli Terry were Ebenezer Parmalee of Guilford, Connecticut, in 1726; Thomas Harland of Boston in 1773; Simon Willard of Grafton, Massachusetts, in 1783; Benjamin Hanks of Litchfield, Connecticut, in 1783; and Gideon Roberts of Bristol, Connecticut, in 1790. Contemporary clockmakers of Terry were, among others, John Rich, Silas Hoadley, Seth Thomas, Chauncey Jerome, Joseph Ives and Luther Goddard. But the one who revolutionized clockmaking in America by being the first to establish a factory for mass-producing low cost clocks was Eli Terry.

Eli Terry was born April 13, 1772, in East Windsor, Connecticut. His father, Samuel, emigrated from England to Springfield, Massachusetts, in 1650. At the age of 14, after only a small amount of common school education, young Terry became an apprentice to clockmakers Daniel Burnap and Timothy Cheney. By 1792, Terry had established himself as a clockmaker in East Windsor, making wood and brass-movement tall-case clocks with brass dials. In September 1793, he moved to Plymouth, Connecticut, started making and repairing clocks, engraving on metal, and selling spectacles. On November 17, 1797, he was granted the first United States patent for an unusual clock, a timepiece showing apparent and mean time on the same dial by two minute hands operating from the same center. In 1800, he decided to increase his production by using water power and became the first clockmaker to use both machinery and water power. Three years later, he was turning out 10 to 20 clocks at a time. This undertaking was the first clock factory in America. In 1806, Terry began making low-priced, wooden 30-hour clock movements for tall-case clocks. In 1807, he obtained a contract with the Reverend Edward Porter and his brother Levi Porter, to make

4,000 wooden clocks at $4 apiece. They, in turn, would attempt to sell these clocks for a profit. To help him in this task, he joined with Seth Thomas and Silas Hoadley to form the firm of Terry, Thomas & Hoadley. After completing the 4,000 clocks in three years, Terry sold out to Thomas & Hoadley in 1810 and established a business of his own at Plymouth Hollow, Connecticut. It was there in 1814 that he devised what he called his perfected wood clock, a 30-hour shelf or mantel clock with a wooden movement and strike to retail for $15. He patented this now-famous pillar scroll top case clock in 1816 and sold the right to make them to Seth Thomas for $1,000. In 1814, Terry induced his sons, Eli, Jr., Henry and Silas to join him in clockmaking at Plymouth Hollow. The firm was known as E. Terry & Sons.

Terry's pillar-and-scroll clocks were so popular that he gradually increased the production of these clocks to 12,000 a year and by 1825 had accumulated a considerable fortune. His clock label, pasted in the backboard of cases, read: "Patent Clocks Invented by Eli Terry. Made and Sold at Plymouth, Connecticut, by E. Terry & Sons." Between the years 1827 and 1833, he gradually withdrew from active clock manufacture but still made brass clocks of fine quality and several tower clocks of novel design. From 1797 to 1845, Eli Terry was awarded nine United States patents for various clocks and improvements. He died at Plymouth, Connecticut, in the part of town known as Terryville, on February 26, 1852.

Color Film

Black and white photography had been around for years when people wanted some color added to their photographs. The two men most responsible for this were Leopold D. Mannes and Leopold Godowsky Jr. They received the first U.S. patent for a commercially acceptable color film. The patent, No. 2,059,884, was granted November 3, 1936.

The history of color films actually begins with the history of black and white silver salt photography. It all started in 1727 when Johann H. Schulze, a German physicist, discovered that silver salts were sensitive to light but that the images produced were not permanent. In 1826, Joseph Nicephore Niepce, a French physicist, found a way to make these silver images permanent. In 1839, William H. F. Talbot, an English scientist, invented the first silver salt negative-positive system of making photographs. In 1871, Richard L. Maddox, an English physicist, introduced gelatin as the vehicle for incorporating silver salts.

White light consists of a mixture of red, green, and blue light. Silver halide is inherently sensitive to blue light but not to red and green. In 1860, James Clerk-Maxwell, a Scottish physicist, showed that all colors may be matched by mixing red, green, and blue light in proper proportions. In 1873, Hermann Wilhelm Vogel, a German chemist, discovered that the addition of certain dyestuffs to a silver halide emulsion conferred sensitivity to green light, and in 1905, Benno Homolko found that silver halide may be sensitized to red light by adding dyes of the carbocyanine class.

In 1887, the Reverend Hannibal Goodwin, of Newark, New Jersey, filed a patent application on a roll film using celluloid as a base for silver halide layers. On November 7, 1888, John Carbutt, of Philadelphia, Pennsylvania, announced to the Photographic Society of Philadelphia the first flexible transparencies using celluloid as a film base. In 1889, Eastman Dry Plate Company of Rochester, New York, was the first to commercially make and market celluloid roll film. In 1897, Louis Ducos du Hauron, a French chemist, patented a film with three superimposed silver halide emulsion layers, sensitized to red, green, and blue light.

While black and white silver halide photographic discoveries were being made, experiments in color photography were also being carried out. The first

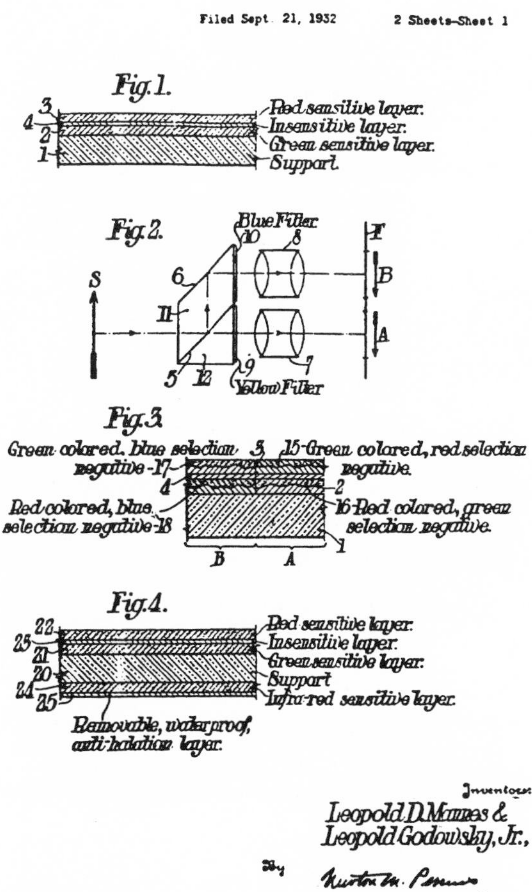

Leopold D. Mannes and Leopold Godowsky Jr. devised a method for color photography that was patented November 3, 1936

color photograph was taken by Thomas Sutton under the direction of Scottish physicist James Clerk-Maxwell and shown at the Royal Institution on May 17, 1861. A color print from the original negatives of this photograph was reproduced in the 1940 *Penrose Annual.* The first color print was displayed before the Societé Francoise de Photographie in Paris on May 7, 1869 and resulted from the work of Louis Ducos du Hauron who patented the subtractive method of color photography on February 23, 1869. Frederick E. Ives of Philadelphia patented the first system of color photography to be developed commercially on February 23, 1890. This system involved making three negatives of the same image at the same time by mirrors in a camera through orange, green, and blue-violet filters. Positives from these negatives were viewed at the same time in a special viewer and illuminated by red, green, and blue-violet light resulting in a single image on a screen in natural color. In 1907, the Lumiere Brothers of France introduced Autochrome color plates that produced photographs in natural color.

A great advance in color photography was made in 1912 when Rudolf Fischer, a German chemist, showed that silver halide emulsions could be developed to color images with dye-coupling developers. In 1921, Dr. L. T. Troland of the United States patented two-layer color monopacks. The first color roll film was introduced in Germany in 1924 and was called Lignose. This was followed by a color roll film called Colorsnap in England in 1929.

The basic problem with all color films up to this point was that of diffusion. The sensitizing dyes and the chemicals that form the dye images would diffuse into adjacent layers, thus eliminating the three-color effect necessary for natural color. This was solved in 1935 by Leopold Mannes and Leopold Godowsky, together with the Kodak Research Laboratories, when Kodachrome was placed on the market. This color film was highly successful and enabled millions of amateur photographers to make color photographs.

Leopold Damrosch Mannes was born in New York City on December 26, 1899. His father, David Mannes, founded and directed the Mannes College of Music in New York. Young Mannes developed an early interest in music, particularly the piano. His other interest was science, and that's what he pursued at Harvard University, graduating in 1920 with a degree in physics. He became a concert pianist and joined the faculty of the Mannes College of Music in 1925, becoming president in 1951. During this time, he toured the United States and abroad as a pianist for the Mannes-Gimpel-Silva Trio. He composed many pieces of music for string quartet, orchestra, and two pianos. He served as a private in the U.S. Army during World War I and did defense work during World War II.

Leopold Godowsky Jr. was born in 1900, in Chicago, Illinois. His father was a noted pianist, composer, and director of the piano department of the Chicago Conservatory of Music. The family moved to Europe where young Godowsky spent much of his youth. It was there that he received musical training, particularly the violin. The family moved to New York City at the outbreak of World War I where young Godowsky enrolled at the Riverdale School in 1916. It was there that he met Leopold Mannes and where both men became interested in color photography. After graduating from the Riverdale School, Godowsky went to California where he worked as a professional violinist with the Los Angeles and the San Francisco symphony orchestras and studied chemistry, physics, and mathematics at the University of California.

In 1920, Godowsky joined his friend, Leopold Mannes, in New York City where they worked from the Mannes kitchen to perfect a color photographic system, supporting their research by working as professional musicians. In 1921, they produced and demonstrated an additive color motion-picture process. This process was not commercial but did attract the attention of a Wall Street company that backed them with $20,000. It also attracted the attention of Dr. C. E. K. Mees, director of the Kodak Research Laboratories, who provided them with equipment and chemicals at cost.

In 1930, Dr. Mees invited Godowsky and Mannes to join the staff of Kodak Research Laboratories in Rochester, New York. There, they developed a three-layer reversal color film in which the three emulsion layers individually produced blue, green, and red negative silver-image records that were reversal processed into corresponding yellow, magenta, and cyan positive dye images. As a result of their work, Eastman Kodak introduced Kodachrome 16 mm motion-picture film in April 1935 and Kodachrome 35 mm and size 828 still films in August 1936.

Godowsky and Mannes remained at Kodak until December 1939. Mannes then became head of the Mannes College of Music in New York City. Godowsky engaged in independent research in color photography in New York and Connecticut under a contract with Eastman Kodak Company. Leopold Mannes died August 11, 1964, in New York City. Leopold Godowsky died February 18, 1983, also in New York City.

Condensed Milk

The first United States patent for a practical method for condensing milk was awarded to Gail Borden Jr. The patent was United States Patent No. 15,553 and was issued on August 19, 1856.

The idea of condensed milk started with Nicholas Appert of France, who in 1804 published a treatise entitled, *The Art of Preserving Animal and Vegetable Substances*. Consequently, Appert is credited with fathering the canning industry. Later, other investigators experimented with methods to preserve milk. These include De Heine of England, who received an English patent for preserving milk and sugar by heating them in an open vessel in 1810; Edward Charles Howard of England, who invented the vacuum pan and received English Patent No. 3,754 in 1813; Malbec of France, who concentrated milk with sugar in 1824; William Underwood of America, who prepared and bottled milk with sugar in 1828; William Newton of England, who received the first English patent for utilizing the vacuum pan in the preparation of milk in 1835; Martin de Lignac of France, who obtained an English patent on evaporated milk preserved with sugar in 1847; C. N. Horsford of England, who prepared condensed milk with the addition of lactose in 1849; Grunaud and Galais of France, who reduced milk to one-fourth its volume at a temperature not exceeding 30 degrees Celsius in 1850; and Grimewade of France, who prepared powdered milk in 1856. However, it was Gail Borden who successfully initiated the commercial manufacture of condensed milk.

Gail Borden Jr. was born November 9, 1801, on his father's farm at Norwich, New York. His mother was the great-great-granddaughter of Roger Williams, who established Rhode Island and founded the Baptist Church in America. His youth was spent on his father's farm, where he was taught the art of surveying. When he was 14, his family moved to Kentucky and then to Jefferson County, Indiana. There in Indiana, Borden obtained his only formal schooling, totaling one and one-half years. While in Indiana, he farmed, practiced surveying, captained a company of militiamen and taught school. Due to ill health, he left home in 1822 for a warmer climate and settled in Amite County, Mississippi. Here, for the next seven years, he taught school in the winter and

15553

Gail Borden Jr. Concentration of Sweet Milk and Extracts. Patented Aug. 19. 1856.

No. 15,553.

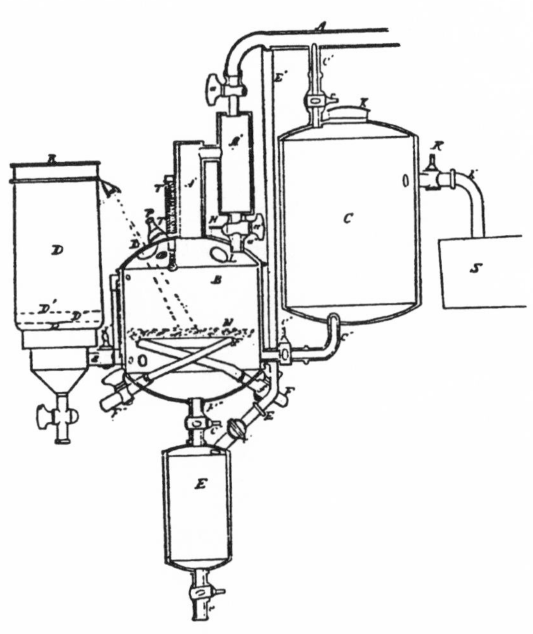

Gail Borden Jr.'s device for the concentration of sweet milk and extracts, patented Aug. 19, 1856

surveyed in the summer. In 1829, Borden and his new bride joined his brother Tom in Stephen A. Austin's colony in Texas, settling on Tom's homestead of 4428 acres to try farming and cattle raising. General Austin appointed him to represent his district at the convention held at San Felipe in 1833 to seek separation from Mexico. During the Mexican War, he and his brother published the *Telegraph and Texas Register*, the only newspaper issued in the territory. When the republic was founded, Borden compiled the first topographical map and laid out the city of Galveston. President Sam Houston made him the Collector of Customs of the Port of Galveston.

Borden now turned his attention to an idea he had long nurtured, that food could be preserved and made safer by condensation. Gold was discovered in California and a party leaving Galveston asked Borden to help them prepare for the journey. Borden set to work to prepare food in concentrated form. He succeeded in preparing a meat biscuit by dehydrating and condensing 120 pounds of beef down to 10, mixing flour with the extract, kneading the substance into biscuits and baking them. The party of Forty-Niners bought 600 pounds of them. Dr. Elisha Kent Kane, the Arctic explorer, used the biscuits on his expedition. Borden exhibited his biscuits at the International Exhibition in London in 1851 and was awarded the Great Council Medal. He was also elected as an honorary member of the London Society of Arts. Borden and his brother built a meat biscuit plant in Galveston. They made every effort to market the biscuit commercially but failed. At the end of five years, Borden's resources were exhausted, and he was heavily in debt.

While returning from the London Exhibition in 1851, he witnessed the death of one infant and the cries of many hungry babies because of the lack of milk. This revived his old interest in inventing a way to preserve milk. Borden left Texas and went to New Lebanon, New York, where he became acquainted with a colony of Shakers, a religious sect at New Lebanon. He began experimenting with milk using one of their vacuum pans of the type used in making sugar. In the spring of 1853, he perfected his technology of condensing milk and in May of 1853, applied for a patent. The Patent Office denied his patent application at first, but Borden was aided by two friends, Robert McFarlane, editor of *Scientific American*, and Dr. John H. Currie, head of an industrial laboratory, who supplied affidavits that the use of the vacuum pan to keep out air during evaporation was a new and important discovery. Due to their help, the three-year fight with the Patent Office ended, and on August 19, 1856, Borden was granted his patent. Two months later, the world's first condensed milk factory opened at Wolcottville, Connecticut. This plant failed but with the help of a newly found friend, wholesale grocer and banker Jeremiah Milbank, a new plant was established at Burrville, Connecticut, as the New York Condensed Milk Company, later to become the Borden Company.

Borden returned to Texas and built a small meat biscuit factory. The little town that grew up around the plant was called Borden, Texas. The inventor lived here until his death on January 11, 1874. On his tombstone in New York's Woodland Cemetery are the words, "I tried and failed, I tried again and again, and succeeded."

Cotton Gin

Eli Whitney, more than any other individual, made it possible for large amounts of cotton to be processed. He was awarded the first United States patent for a cotton gin on March 14, 1794. This was the 72nd patent issued by the recently established United States.

Before Eli Whitney, there is no history concerning the cotton gin. It appears to be an American invention by a single American. The roller gins, adapted to crush the seeds rather than separate them from the fiber, and the Indian churkas, used for cleaning cotton, were the only other methods, besides hand-picking, used in the treatment of cotton. The popularity of the Whitney cotton gin is attested by the many applications for cotton gin patents that came shortly after the Whitney invention. Some of these inventors were Hodgen Holmes in 1796, Robert Watkins in 1796, John Murray in 1796, Eben Whiting in 1801, William Bell in 1802, J. S. D. Montmollen in 1802, G. F. Saltonstall in 1803, and John McBride in 1805. But it was Eli Whitney who single-handedly revolutionized the cotton industry with the invention of the first cotton gin.

Eli Whitney was born at Westboro, Massachusetts, on December 8, 1765. His father was a prosperous farmer, but young Whitney showed more interest in his father's shop, fitted with a variety of tools and a turning lathe, than he did in farming. At the age of 12, he made and repaired violins and earned local popularity by playing at country dances. Before long, the neighbors kept him busy repairing anything from violins to broken chairs. During the Revolutionary War, when he was only 16, he manufactured nails from a nail machine that he made. After the war, he turned to making hatpins so skillfully that he almost monopolized that business.

When he was 19, Whitney decided that he needed more education. To obtain funds, he applied for a position as schoolmaster in nearby Grafton. Although not qualified, he was selected to teach and with the money he earned attended Leicester Academy in Leicester, Massachusetts. In 1789, he entered Yale College at the age of 23 and graduated in 1792, earning a considerable portion of his school expenses by repairing articles around the college. Upon graduation, he decided to study law. He accepted a position as tutor with a Georgia family so he might have funds and leisure time to study law. On the boat to Savannah, Georgia, Whitney became acquainted with Catherine Greene, widow of General Nathanael

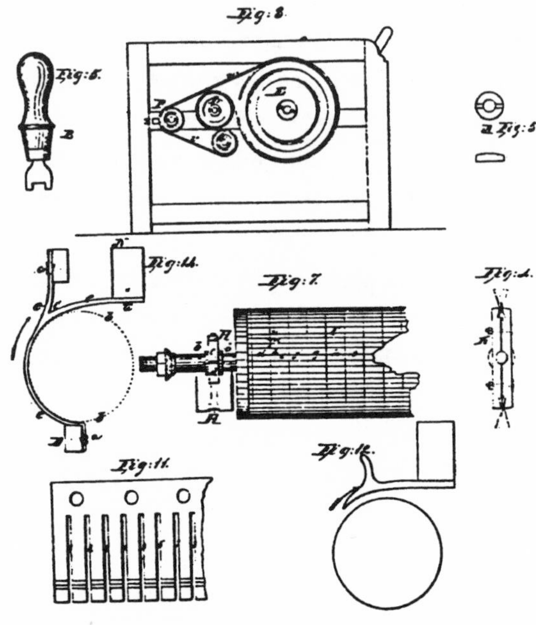

Eli Whitney revolutionized the the cotton industry with the invention of the first cotton gin, patented on March 14, 1794

Greene, a hero of the Revolutionary War, and Phineas Miller, the manager of her plantation. Mrs. Greene offered him the hospitality of her Mulberry Grove estate on the Savannah River after Whitney learned that his prospective employer had engaged another tutor. Whitney gratefully accepted and began his law studies. However, he found that most of his time was occupied with repair jobs on the plantation.

In 1792, a group of southern planters were entertained at the Greene plantation. During the evening, the conversation focused on the distressing conditions of the southern cotton industry. One of the gentlemen remarked that the cotton troubles would be eliminated if some machine could be devised to separate the green seed from the cotton fiber. Mrs. Greene introduced them to Whitney and stated that he could make such a machine. With encouragement from Mrs. Greene and Miller, he set to work and six months later produced a machine that could do the work of 50 laborers.

Whitney and Miller formed a partnership in 1793, expecting to monopolize the southern cotton industry by purchasing, cleaning, and selling the entire output. This scheme failed, because the device was so simple to manufacture and the principle so easy to copy that, before he received his patent in 1794, models of his gin were in use throughout the South. Whitney went to New Haven, Connecticut, and opened a factory. There, he planned to build gins and lease them to the planters on a 33 1/3 percent royalty basis. These plans also met with failure as his buildings and machinery were destroyed by fire in 1795. He had to contend for years with many obstacles: poverty, sickness, infringement of his patent, over 60 law suits, the burning of his factory, and the death of his partner were just a few. The term of his patent had nearly expired before he obtained a single decision of a court in his favor. Only three states, South Carolina, North Carolina and Tennessee, agreed to buy the rights to his invention. Even Congress voted not to extend the length of time on his patent.

Seeing that there was little prospect of ever receiving much remuneration from his invention of the cotton gin, Whitney turned his attention to other interests. He learned that the government in May of 1798 had voted $800,000 for the purchase of firearms. He made a successful bid for part of this business, although he had no experience in this field. He contracted with the government, through the influence of Secretary to the Treasury Oliver Wolcott, to make 10,000 muskets in 28 months. This contract with the government was unique at the time in that it allowed an advance of $5,000 and the promise of another $5,000 as soon as the first was spent. Whitney said later that this feature saved him from bankruptcy and ruin. He purchased land at Mill Rock, Connecticut, later called Whitneyville, put up buildings, designed and built machines, recruited and trained workmen, and started production. The whole contract took over 10 years to complete, but it was there at Whitney's Mill Rock factory that the system of using interchangeable components for large-scale manufacture was installed and achieved widespread acceptance. It was there also that he established the first "company town" in America.

Whitney had other inventions. He was the inventor of the first milling machine. He also invented a jig, a screw press, a belt-driven trip hammer, and a machine for boring wood. However, after his bitter struggles and numerous law suits over the patent rights for the cotton gin, he not only refused to patent, but exhibited and discussed freely, all his later inventions. He was one of the charter members of the Hall of Fame for Great Americans when it was established in 1900. Whitney died at New Haven, Connecticut, on January 8, 1825.

Cyclotron

The first United States patent for a scientific instrument to disintegrate atomic nuclei to form radioactive substances was awarded to Ernest Orlando Lawrence. The instrument was called a cyclotron and was awarded United States Patent No. 1,948,384 on February 20, 1934.

Many people have contributed to the study of the atom. These include the work in 1897 of British physicist Joseph John Thomson, who showed that cathode rays were particles of high velocity that could be deflected with an electromagnet; French chemists Pierre and Marie Curie, who discovered radioactivity in 1898; British physicist Ernest Rutherford, who postulated that the atom was made up of a positive nucleus surrounded by a sphere of electrons in 1911; Scottish physicist Charles T. R. Wilson, who discovered the cloud chamber in 1911; Danish physicist Neils H. D. Bohr, who in 1913 concluded that the basic atomic unit must be the hydrogen atom of mass 1 and advanced the idea that electrons sent out their energy in packets or quanta, and that an electron moving from one orbit to another emitted a burst of this energy; British physicist Henry G. J. Mosely, who in 1914 showed that each element when excited by X-rays, emitted waves characteristic of its atomic number; Austrian physicist Erwin Schrodinger, who in 1926 founded the science of wave mechanics; British physicist John Douglas Cockcroft and Irish physicist Ernest Thomas Walton, who invented the linear particle accelerator in 1929; American chemist Harold C. Urey, who made heavy hydrogen in 1931; British physicist James Chadwick, who discovered the neutron in 1932; and American physicist Charles D. Anderson, who discovered the positron in 1932. Ernest O. Lawrence, an American physicist, added to this knowledge by building an accelerator of sufficient strength, called the cyclotron, by running protons through alternating electric fields in a spiral path in 1934.

Ernest Orlando Lawrence was born in Canton, South Dakota, on August 8, 1901. His father, Carl G. Lawrence, was president of Northern State Teachers College in Aberdeen, South Dakota. He obtained his elementary education in the public schools of Canton and Pierre, South Dakota. He attended college first at St. Olaf College and then at the University of South Dakota. He did graduate work at the University of Minnesota and the University of Chicago before obtaining his PhD degree in 1925 at Yale University.

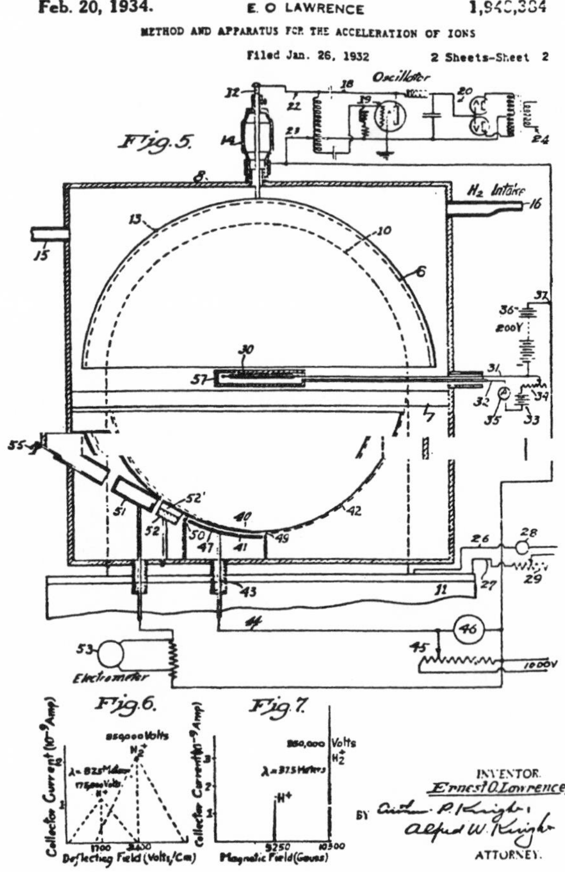

E. O. Lawrence received a patent on the method and apparatus for the acceleration of ions on February 20, 1934

He remained at Yale for two years, first as a National Research Fellow and then as assistant professor. In 1927, he joined the faculty of the University of California and remained there the rest of his career.

In 1929, Lawrence turned his attention to the problem of accelerating atomic particles after reading a paper by the German physicist R. Wideroe. In 1930, he and a graduate student, N. E. Edlefsen, built the first cyclotron. This was a small device, only four inches in diameter and constructed of glass and red sealing wax, in which protons were made to travel between the poles of a large magnet that deflected their paths into circles. At each turn, they received another small push of electric potential that made the protons move faster and faster in a less sharp curve. Their path was in the form of a spiral that brought them closer and closer to the rim of the instrument. When the charged particles finally shot out of the instrument, they had accumulated very high energies. These high energy particles were then directed at various targets. Larger metal cyclotrons were built by Lawrence, including one with a 60-inch diameter and a magnet weighing 180 tons that he constructed in 1939. Machines of this size easily make quantities of artificial radioactive substances for medical use. During World War II, Lawrence used the cyclotron to separate quantities of uranium-235 from ordinary uranium to be used in Fermi's atomic pile in Chicago.

Many awards were presented to Lawrence. He was elected to membership in the National Academy of Sciences in 1934, the first native of South Dakota to receive this honor. In 1937, he was presented the Comstock Prize of the National Academy of Sciences. He was awarded the 1939 Nobel Prize in physics and the Enrico Fermi Award of the Atomic Energy Commission in 1957. Lawrence died on August 27, 1958, at Palo Alto, California. In 1961, element 103 was discovered and was named Lawrencium in his honor.

Design Patent

The first United States design patent was issued to George Bruce of New York City on November 9, 1842, for four new and different printing type faces.

Prior to 1842, there were no laws in the United States affording protection to designs for articles of manufacture. Design statutory protection in European nations had been in effect for many years before similar laws were enacted in the United States. Protection was available in France as early as 1737 and in England as early as 1787. Commissioner of Patents Henry L. Ellsworth, in his report to the 27th Congress, 2nd Session, of February 8, 1841, called attention to the lack of protection for new and original designs and suggested the passage of such an act. He stated, "It may well be asked if authors can so readily find protection in their labors, and inventors of the mechanical arts so easily secure a patent to reward their efforts, why should not discoverers of designs, the labor and expenditure of which may be far greater, have equal privileges afforded them?" As a result of the recommendation of Commissioner Ellsworth, Congress enacted the first Design Law on August 29, 1842. The first inventor to take advantage of this new law was George Bruce of New York City on November 9, 1842, who was awarded United States Design Patent No. 1. Under the 1842 act, 1,277 design patents were issued.

George Bruce was born in Edinburgh, Scotland, on June 26, 1781. When he was 14 years old, he emigrated to the United States, arriving in Philadelphia in June 1795. His brother David had emigrated to New York City two years earlier in the spring of 1793. George served three years in Philadelphia as an apprentice to Thomas Dobson, a printer. He moved to New York in 1798 and worked as a journeyman until 1803, when he became foreman and occasional contributor to the *New York Daily Advertiser*. In 1806, he, with brother David, established a book printing office, known as D & G Bruce, Printers. One of the first books they printed was *Lavoiser's Chemistry*.

In 1812, they established Bruce's New York Type-Foundry, and in the same year established a stereotype-foundry, the first in America. The type-founding business increased so much that the brothers finally quit printing and made letter-casting and stereotyping their main business. They were the first in America to stereotype the New Testament and Bible. In 1822, the company was dissolved due to the retirement of David to the farm. His son, David

The first United States design patent, issued in 1842 to George Bruce of New York City for four new and different printing type faces

Bruce Jr., continued in the printing business and patented the first type-casting machine in 1836.

George continued to work in the type-founding business. He harmonized and graduated the size of the eleven series of the different bodies of type, and introduced the Agate body for the first time in the series. He was also distinguished for his skill as a punch-cutter, particularly Romans and scripts. In 1854, he patented a method of cooling the mold of the type-casting machine invented by his nephew, David Bruce Jr. In 1863, he was elected president of the Type-Founders Association, a position he held until his death. He died July 5, 1866, in New York City at the age of 85.

Diesel Engine

The first U.S. patent for a diesel engine was awarded to Rudolf Diesel on August 9, 1898. The patent number was 608,845.

Many people previous to Diesel contributed to the development of the internal combustion engine. These include Samuel Morey in 1826, Etienne Lenoir in 1860, Alphonse Beau de Rochas in 1860, George Brayton in 1872, Nicolaus Otto in 1877, Gottlieb Daimler in 1884, Herbert Akroyd-Stuart in 1885, and Karl Benz in 1885. However, it was Rudolf Diesel, who, in 1892, began to experiment with an internal combustion engine that would later profoundly affect the world of transportation.

Rudolf Diesel was born in Paris of German parents on March 18, 1858. The boy grew up sharing his parent's fluency in English, French, and German. In 1870, the family moved to London, having been forced to leave France because of the Franco-Prussian War. Young Diesel, at age 12, was sent to live with an uncle in Augsburg, Bavaria. While living there, he graduated from the Industrie Schule, an engineering school, and earned a scholarship at Munich Polytechnic, a technical college in Munich. While at this college, he heard Professor von Linde, head of the college and the inventor of refrigeration machines, give a lecture on the inefficiency of the steam engine. From the lecture, Diesel learned that steam engines transform only 12 percent of the supplied heat into energy. He made a note in the margin of his notebook about possibly getting more energy from the heat. This thought stayed with him for many years. He graduated from Munich Polytechnic at the age of 21 and became assistant to Professor von Linde. Later, he took a position with von Linde's refrigeration firm and built a plant in Paris, where he remained for 10 years. He then took an assignment to handle von Linde's business in North Germany.

While working for von Linde, Diesel had ample opportunity to pursue his dream of making a more efficient internal combustion engine. His first efforts centered around an ammonia vapor motor, and he actually gave a paper on the subject before the 1889 International Congress of Applied Mechanics in Paris. Soon he turned his attention to the four-cycle internal combustion engine developed by Nikolaus Otto and designed an engine that did not require an electric ignition system, and which used a mixture of oil and air rather than gasoline and air. On February 28, 1892, the German Patent Office granted Diesel

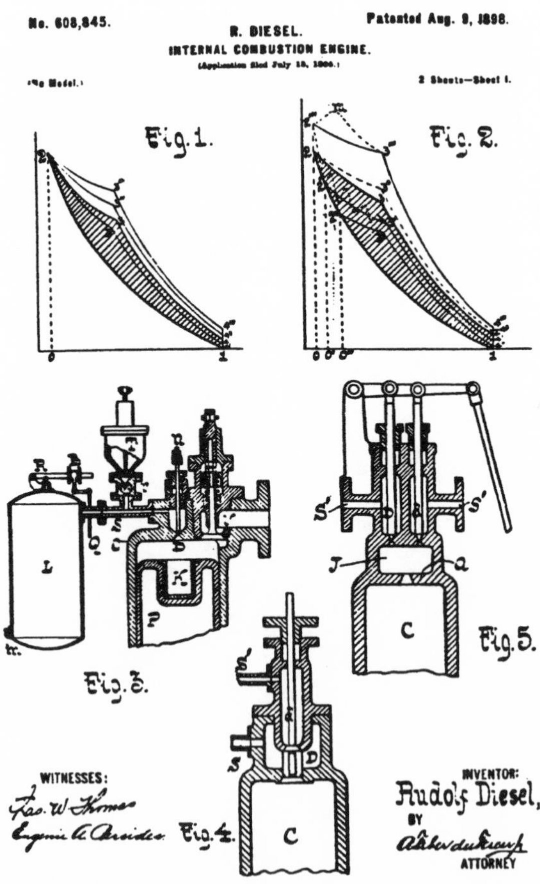

Rudolf Diesel's internal combustion engine, patented on August 9, 1898

patent No. 67,207 for his engine. In 1893, he published a book, *Theory and Construction of a Rational Heat Engine to Replace Steam and the Existing Internal Combustion Engine*. This book, together with Diesel's patent, convinced the Augsburg Machine Works and the Krupp Works to pay the inventor 30,000 marks annually to construct his engine. On August 10, 1893, he tested the first diesel engine. It exploded but the test proved that heat resulting from compression would ignite the fuel. He then designed and built a new engine over the next two years, and in June 1895, this engine tested satisfactorily. His fame soon spread through the engineering world.

The diesel engine was greeted with enthusiasm everywhere. The engine needed neither spark plugs nor an ignition system. It did not require a carburetor to change the liquid fuel into gas and mix it with air. It could use cheap, heavy oil instead of the more expensive gasoline. The efficiency was greater than any other internal combustion engine. Diesel founded a world-wide company with headquarters in Munich and traveled to all parts of the world to license his engine. In America, the first diesel engine built for commercial service was a two-cylinder 60 H.P. unit built in September 1898 in the plant of the St. Louis Iron and Marine Works. This engine, which drove a direct-current generator, was erected and operated in the Second Street plant of the Anheuser Busch brewery. Adolphus Busch bought Diesel's American patent rights in 1897 for a sum of approximately $250,000 and established the Diesel Motor Company of America.

On October 1, 1913, there was to be a meeting in London of the Consolidated Diesel Manufacturing Limited. On September 29, 1913, Diesel boarded a ship in Antwerp, Belgium, to attend. He never reached England. During the night, he mysteriously disappeared, and the mystery was never solved. Later, when his papers were examined, it was found that his financial condition was desperate. He had embarked on all kinds of losing business transactions, including oil and land interests, and 1.25 million marks of liabilities were found to be without securities. Diesel never lived to see that his engine had revolutionized marine, highway, and railway transportation.

Dirigible

The first United States patent for a dirigible was awarded to Count Ferdinand von Zeppelin. The patent number was 621,195, issued on March 14, 1899.

France and Germany can claim credit for the creation of the airship. France produced enthusiasts like Etienne, Joseph and Charles Mongollier; Henri Albert Julliot; Jean Baptiste Meusnier; Paul and Pierre Lebaudy; Henri Giffard; Gaston and Albert Tissandier; Charles Renard; Authur Krebs; and Alberto Santos-Dumont, a Brazilian who worked in France. From Germany came enthusiasts like Paul Haenlein, Hermann Ganswindt, August von Parseval, Dr. Wolfert, and David Schwartz. However, credit for designing, building, and flying the first reliable dirigible airship belongs to Zeppelin.

Count Ferdinand von Zeppelin was born in Konstanz, Germany, on July 8, 1838. His father, Count Frederick Zeppelin, was a German aristocrat. Upon graduation from various military schools, young Zeppelin entered the army and was commissioned as a lieutenant in 1857. He was sent to the United States in 1863 as an observer with the Northern Army of the Potomac during the Civil War. While in the United States, he experienced his first balloon ascension in St. Paul, Minnesota. He returned to Germany to take part in the Austro-Prussian War of 1866 and the Franco-Prussian War of 1870. He retired from the German Army in 1891 with the rank of lieutenant general.

Having ample time and money, he turned his attention to flying. He employed an engineer, Theodore Kober, to design an airship. The famous cigar-shaped drawings were completed in 1894 and submitted to the German Military Commission. The Commission subsequently turned the drawings down on the grounds that such a machine would be too weak and too slow. Zeppelin then founded the Zep-

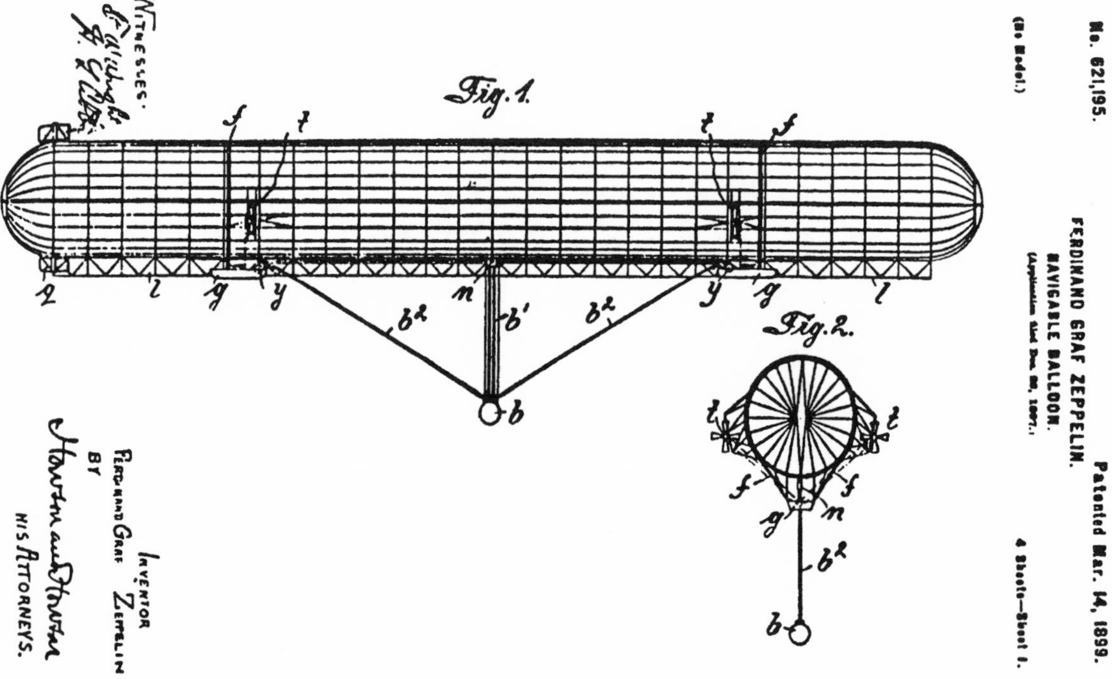

Count Ferdinand Zeppelin's navigable balloon, patented March 14, 1899

pelin Airship Building Company in 1898, built a floating shed on Lake Konstanz at Manzell, Germany, and constructed the first Zeppelin airship, the LZ-1, in 1900. The airship weighed 25,350 pounds with a capacity of 400,000 cubic feet of gas and was 420 feet in length. It was powered by two Daimler 16 H.P. engines. The first flight on July 2, 1900, with Zeppelin at the controls, resulted in the airship being damaged on landing. It was, however, the first effective directed flight by man.

Count Zeppelin was determined to build another airship but had trouble getting financial backing until the State of Wurttemberg gave him permission to run a lottery to raise funds. With these funds and money from a wealthy aluminum manufacturer named Herr Berg, Zeppelin built the LZ-2, which made its first flight on November 30, 1905. This flight also resulted in damage to the airship due to a wind-storm.

With help from his friends, Zeppelin built the LZ-3. He had improved this airship by employing 85 H.P. engines and an improved steering method. The maiden flight was on October 9, 1906, and this time, it was completely successful. The government at last took an interest and awarded him a million marks for further developmental work. The German Emperor also awarded Zeppelin with the Order of the Black Eagle.

Although the first airships were built for the army, it was decided that they also could be used to carry passengers. So the German Airship Transport Company was formed at Frankfurt on November 16, 1909, and by 1913 was making flights reliably and punctually with Hugo Eckener as captain. World War I put a temporary end to passenger service but not before 35,000 passengers traveled 170,000 miles with no loss of life.

Zeppelin died on March 8, 1917, at Charlottenburg, Prussia. He did not live long enough to see one of his airships, the *Graf Zeppelin,* with Hugo Eckener as commander, become the first lighter-than-air craft to circle the globe in 1929. Nor did he live long enough to see the beginning of the end to the development of lighter-than-air transport. The end came in 1937 when Germany's largest Zeppelin, the *Hindenburg*, exploded and burned as it tried to land at Lakehurst, New Jersey, with the loss of 36 lives. Before this disaster, however, the airship had made 56 flights, 10 of them to North America.

Dynamite

Nitroglycerine was discovered by Ascanio Sobrero in 1847. Twenty-one years later Alfred Nobel mixed it with kieselguhr and received the first United States patent for dynamite, No. 78,317, on May 26, 1868.

The history of explosives is old and included many people before Nobel came on the scene. Some of the more prominent ones are Berthold Schwartz of Germany who invented gunpowder in 1320; David Ramsay of England who discovered saltpeter in 1630; Alexander Forsythe of Scotland who made percussion caps in 1805; William Congreve of England who invented the time-fuse in 1823; Moses Shaw of the United States who received a patent for a method of setting off numerous blasts of gunpowder simultaneously in 1830; Henri Braccanot of France who discovered xyloidine in 1832; Theophile Jules Pelouze of France who made nitrostarch in 1838; Christian Schönbein of Switzerland who invented guncotton in 1846; Ascanio Sobrero of Italy who discovered nitroglycerine in 1847; and Tal P. Schaffner of the United States who made nitroleum in 1866. However, it remained for Alfred Nobel of Sweden in 1868 to invent the most promising explosive—dynamite.

Alfred Bernhard Nobel was born in Stockholm, Sweden, on October 21, 1833. His father, Immanuel Nobel, invented a submarine torpedo that caught the attention of the Russian ambassador to Sweden. The Russian government invited him to build and operate a torpedo factory for the Czar in St. Petersburg, Russia. The family moved to Russia in 1842, when Nobel was 10 years of age. Poor health due to a spinal ailment prevented him from attending school, so most of his instruction came from his mother and other tutors. He learned to speak and write Russian fluently and later mastered French and German. When he was age 15, Nobel became an apprentice in his father's machine shop. In 1850, his father sent him for travel and study to France, Italy, Germany, and finally to the United States to be trained as a mechanical engineer under John Ericsson, designer of the *Monitor*. While in America, he received a thorough engineering training and mastered the English language.

In 1854, Nobel rejoined his family in Russia where his father was engaged in the manufacture of explosives. There, the inventor busied himself in his father's shop and received patents on a new gasometer, an apparatus for measuring liquids, and a barometer—the first of some 355 patents he would receive during his lifetime. He and his father began

United States Patent Office.

ALFRED NOBEL, OF HAMBURG, GERMANY, ASSIGNOR TO JULIUS BANDMANN, OF SAN FRANCISCO, CALIFORNIA.

Letters Patent No. 78,317, dated May 26, 1868.

IMPROVED EXPLOSIVE COMPOUND.

The Schedule referred to in these Letters Patent and making part of the same.

TO ALL WHOM IT MAY CONCERN:

Be it known that I, ALFRED NOBEL, of the city of Hamburg, Germany, have invented a new and useful Composition of Matter, to wit, an Explosive Powder.

The nature of the invention consists in forming out of two ingredients long known, viz, the explosive substance nitro-glycerine, and an inexplosive porous substance, hereafter specified, a composition which, without losing the great explosive power of nitro-glycerine, is very much altered as to its explosive and other properties, being far more safe and convenient for transportation, storage, and use, than nitro-glycerine.

In general terms, my invention consists in mixing with nitro-glycerine a substance which possesses a very great absorbent capacity, and which, at the same time, is free from any quality which will decompose, destroy, or injure the nitro-glycerine, or its explosiveness.

It is undoubtedly true, as a general rule, that nitro-glycerine, when mixed with another substance, possesses less concentration of power than when used alone; but while the safety of the miner (to prevent leakage into seams in the rock) prohibits the use of nitro-glycerine without cartridges, which latter must of course be somewhat less in diameter than the bore-holes which are to contain them, the powder herein described can be made to form a semi-pasty mass, which yield[illegible] the slightest pressure, and thus can be made to fill up the bore-hole entirely. Practically, therefore, the miner will have as much nitro-glycerine in the same height of bore-hole with this powder as with nitro-glycerine in its pure state.

This is the real character and purpose of my invention; and in order to enable others skilled in the art to which [illegible] appertains (or with which it is most nearly connected) to make, compound, and use the same, I will proceed to describe the same, and also the manner and process of making, compounding, and using it, in full, clear, and exact terms.

The substance which most fully meets the requirements above mentioned, so far as I know or have been able to ascertain from numerous experiments, is a certain kind of silicious earth or silicic acid, found in various parts of the globe, and known under the several names of silicious marl, tripoli, rotten-stone, &c. The particular variety of this material which is best for my compound is homogeneous, has a low specific gravity, great absorbent capacity, and is generally composed of the remains of *infusoria*.

So great is the absorbent capacity of this earth, that it will take up about three times its own weight of nitro-glycerine and still retain its powder-form, thus leaving the nitro-glycerine so compact and concentrated as to have very nearly its original explosive power; whereas, if another substance, having a less absorbent capacity, is used, a correspondingly less proportion of nitro-glycerine will be absorbed, and the powder be correspondingly weak or wholly inexplosive.

For example, most chalk will take but about fifteen per cent. of nitro-glycerine and retain its powder-form. Twenty per cent. will reduce it to a paste.

Porous charcoal has also a considerable absorbent capacity, but it has the defect of being itself a combustible material, and also of less elasticity of its particles, which renders it easy to squeeze out a part of its nitro-glycerine.

The two materials are combined in the following manner:

The earth, thoroughly dried and pulverized, is placed in a wooden vessel. To it is introduced the nitro-glycerine in a steady stream so small that the two ingredients can be kept thoroughly mixed.

The mixing may be effected by the naked hand, or by any proper wooden instrument used in the hand, or by wooden machinery

Sufficient of nitro-glycerine should be used to render the compound explosive, but not so much as to change its form of powder to a liquid or pasty consistency.

Practically, about sixty parts, by weight, of nitro-glycerine [illegible] forty of earth, forms the useful minimum,

Dynamite, a discovery by Alfred Nobel, was patented on May 26, 1868

to work with nitroglycerine, which had been discovered a decade earlier by Ascanio Sobrero, an Italian professor at the University of Turin. At this time, it was called Nobel's Blasting Oil.

After losing the Crimean War, the Czar lost interest in torpedoes and the Nobel family. In 1859, the Nobel family moved back to Stockholm and continued to experiment with nitroglycerine, which by this time had shown great promise as an explosive. However, Swedish bankers would not invest in a Nobel nitroglycerine factory that had the potential of blowing up at any time. Nobel was sent to Paris to raise money in 1861 and, through the influence of Napoleon III, raised 100,000 francs from the Paris banker Peroire. This money was used to build a small factory at Heleneborg, near Stockholm, in 1862. Things went well until September of 1864, when Heleneborg was rocked by an explosion. The blasting oil plant had blown up, killing Nobel's brother Emil. The Swedish government soon notified Nobel that no more experimentation with explosives could be done near populated areas. So Nobel rented a barge and moored it in the middle of Lake Malaren and continued his experiments there.

While on this barge, Nobel began work on a method to tame nitroglycerine. He made two discoveries that accomplished this feat. He discovered that nitroglycerine when mixed with diatomaceous earth, kieselguhr, could be handled safely and yet, when set off, would retain its powerful explosiveness. He also discovered that this combination could be set off with a blasting cap of mercury fulminate. With these two discoveries, he had no problem finding financial backers in many countries. In 1865, he founded the firm of Alfred Nobel & Company and erected factories over the next few years in Germany, Sweden, Norway, United States, Finland, Scotland, France, Spain, Switzerland, Italy, Portugal, and Hungary.

Nobel amassed an immense fortune from explosives and from investments in the Baku oil wells in Russia. He did not want to be remembered as "a merchant of death," which was how he was described in a mistaken and premature obituary, so he established and funded the Nobel Prizes. At his death he left a fund of $9,200,000 for the establishment of annual prizes in the fields of peace, literature, physics, chemistry, and medicine. The first prizes, awarded in 1901, carried a cash value of $30,000 each. The Nobel Peace Prize was awarded to Henri Dunant of Switzerland and Frederick Passy of France; the Nobel Prize in literature was awarded to Rene F.A. Sully Prudhomme of France; the Nobel Prize in physics was awarded to Wilhelm K. Rontgen of Germany; the Nobel Prize in chemistry was awarded to Jacobus H. Van't Hoff of the Netherlands; and the Nobel Prize in medicine was awarded to Emil A. von Behring of Germany. The Nobel Memorial Prize in Economic Science was added later in 1968.

Alfred Bernhard Nobel died of a cerebral hemorrhage in San Remo, Italy, on December 10, 1896.

Electric Motor

One cannot look in any direction, whether at home or work, without seeing an appliance or machine that operates by an electric motor. Although most of these are alternating-current electric motors, it was the direct-current electric motor that was a precursor, and it all started with a man named Thomas Davenport. He secured the first U.S. patent for an electric motor. The number was 132, issued February 25, 1837.

The invention of the electric motor was made possible by the invention of the electromagnet by Joseph Henry. Contemporaries of Davenport were working on the electric motor at about the same time. These inventions include Joseph Henry's motor of 1830, Sturgeon's rotary motor of 1832, Edmondson's motor of 1834, Watkins's rotary motor of 1834, Jacobi's rotary motor of 1834, Zabriskie's motor of 1837, Slade's rotary motor of 1837, Page's rotary motor of 1838, Wakley's motor of 1838, Stimson's motor of 1838, Page's motor of 1839, Cook's motor of 1840, and Elias's motor of 1842. However, it was Davenport who received the first U.S. patent for his direct-current electric motor on February 25, 1837.

Thomas Davenport was born in Williamstown, Orange County, Vermont, on July 9, 1802. The eighth of 11 children, he had attended public school for only a few years when his father died, and he was forced to stop and begin working to help support his family. When he was 14 years old, he was apprenticed to a local blacksmith and remained in this position for seven years. During these seven years, he saved as much money as he could with the idea of opening his own business. Finally, in 1823, he opened his own blacksmith shop in Brandon, Vermont. There, he prospered, married a local girl in 1827, and built a large brick house.

In 1831, a Joseph Henry electromagnet was put to use at the Penfield Iron Works at Crown Point, New York, for sifting and separating magnetic ore. Word of this electromagnet spread across the Vermont line to Brandon and finally to Thomas Davenport. He borrowed his brother's horse to go to Crown

T. DAVENPORT.
ELECTRICAL MOTOR.

No. 132. Patented Feb. 25, 1837.

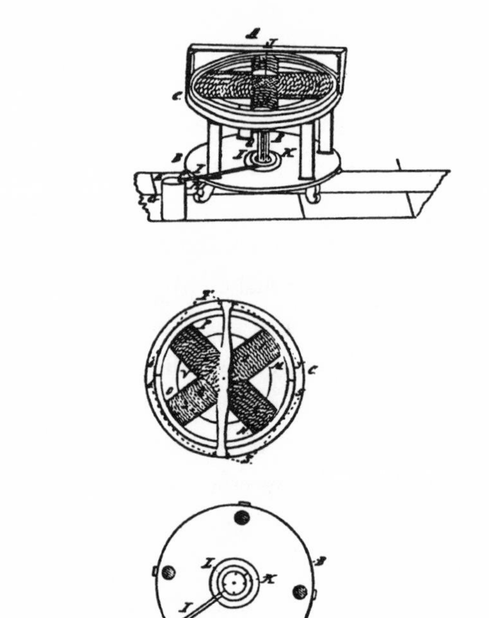

Witnesses. Inventor.

The patent for the direct-current electric motor, granted to Thomas Davenport on February 25, 1837

Point to buy some iron for his blacksmith shop, but when he saw the electromagnet at work, he became so fascinated that he bought the electromagnet instead of the iron. He even traded his brother's horse for another inferior animal and some cash, and with this cash bought a second electromagnet. With these two electromagnets, Davenport studied and constructed one of his own, using strips of his wife's silk wedding dress as insulation.

He was then struck with the idea that this device could be modified to produce a rotary motion and become a source of power. Neglecting his blacksmith duties, he began to devote more and more time to this idea. After several months and hundreds of experiments, he finally succeeded in July of 1834 in building the first practical electric motor. It consisted of a wheel that contained two electromagnet spokes. He mounted the wheel so it would revolve in a horizontal plane and placed two fixed electromagnets in the same plane with their poles pointed toward the spokes. He added a commutator and brushes of sorts and hooked the device to a battery. By repeatedly changing the direction of the current, he caused the magnets to repel and attract one another alternately so the wheel revolved at an extremely low speed. With the help of an anonymous, wealthy backer, Davenport improved the motor so it would rotate about 30 revolutions per minute.

In December 1834, Davenport exhibited his motor to a college professor at Middlebury College in Middlebury, Vermont. This professor encouraged him to the extent that he exhibited his machine to Professor Amos Eaton of the Rensselaer Polytechnic Institute in Troy, New York; to Professor Joseph Henry at Princeton, New Jersey; and to Professors Bache and Hamilton at the Franklin Institute in Philadelphia. Being encouraged by these men of science, he traveled to Washington, D.C. in 1835 and applied to the United States Patent Office for a patent. In the summer of 1836, he met Ransom Cook, a mechanic of Saratoga Springs, New York, and together they built a model and deposited it with the Patent Office. However, on December 15, 1836, a fire at the Patent Office destroyed both the application and the model. A second application and model were immediately submitted and Davenport received U.S. Patent No. 132 on February 25, 1837.

For the next six years, he tried, without success, to secure financial backing to commercialize his invention. He even put the invention into a Joint Stock Association of 3,000 shares, said to be the first electric stock company in America, if not the world. Enough stock was sold to secure patents in England, Ireland, and Scotland but not enough to insure the success of the invention.

Davenport made other important advances in the scientific world. He built what is now recognized as the forerunner of the electric trolley car. He also built the first printing press operated by electricity using one of his own motors. On this printing press he published a technical journal called the *Electro-Magnet and Mechanics Intelligencer.*

In 1843, while living in New York City, he broke down physically and returned to Brandon, where he built an electromagnetic player piano. Three years later, he retired to a small farm in Salisbury, Vermont. He died July 6, 1851.

Electric Welder

Although his name is not generally known, Elihu Thomson contributed to the progress of better living by inventing electric resistance welding, a process by which pieces of metal are joined permanently together by passing an electric current through the joint. For this invention, he was awarded the first U.S. patent for electric welding—No. 347,140, issued August 10, 1886.

All electric welding can be divided into two general classes: arc welding and resistance welding. Arc welding is the older of the two processes. Credit for modern arc welding is given to Sir Humphrey Davy who in 1801 discovered that an arc could be created between two terminals of an electric circuit of high voltage by bringing them near each other. The first attempt to use the intense heat of the carbon arc for welding purposes was made in 1881 by de Meritens who joined parts of a storage battery plate by that process. Others who worked on arc welding were Zerner, Bernardos, Slavianoff, Strohmenger, and Slaughter. However, the idea of joining metals by an electric current, known as the resistance method and corresponding apparatus, was pioneered by Elihu Thomson.

Elihu Thomson was born in Manchester, England, on March 29, 1853. Five years later, the family moved to Philadelphia, Pennsylvania. He finished the elementary education courses in the Philadelphia public schools at age 11, but had to wait two years before being admitted to Philadelphia's Central High School where he was an outstanding student and graduated in 1870 with high honors. That same year, he was invited to join the faculty of Central as assistant professor of chemistry. In 1876, he was given the chair of chemistry at age 23, a position he retained until 1880. His interest in electricity was enhanced by his friendship with Edwin J. Houston, professor of natural philosophy at the high school. They worked together to produce, in 1879, the Thomson-Houston system of arc lighting.

In 1880, he gave up teaching to devote his time exclusively to developing and manufacturing electrical apparatus. He was called to New Britain, Connecticut, to become the electrical engineer of a new company that was being organized to develop the Thomson-Houston patents, called the American Electrical Company. There, Thomson further perfected the arc-lighting system, improved the dynamo, and developed an automatic current regulator. In 1883, the company moved to Lynn, Massachusetts, changed

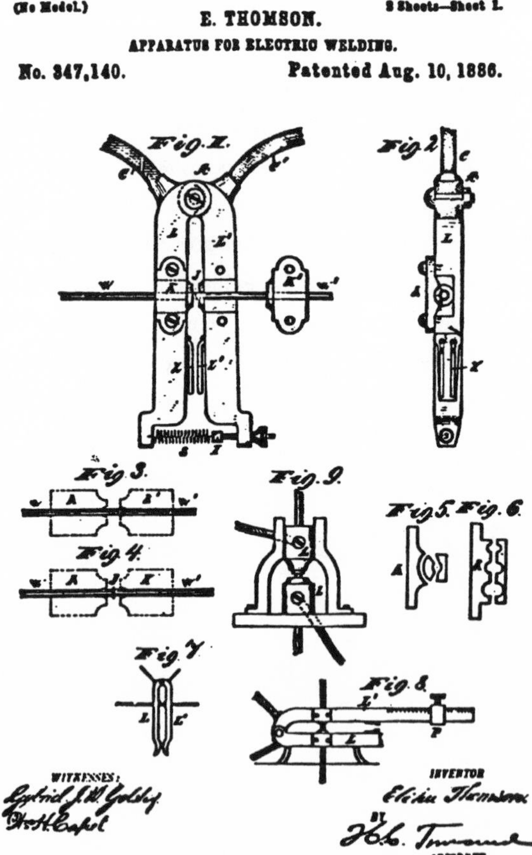

The invention of electric resistance welding is credited to Elihu Thomson in this patent granted August 10, 1886

its name to the Thomson-Houston Electric Company. Thomson was retained as chief electrical engineer. This company later merged with the Edison General Electric Company, and is now the General Electric Company. It was there that Thomson invented the wattmeter, the grounded transformer, the oil-cooled transformer, and the resistance method and apparatus of electric welding.

Thomson's resistance method of electric welding occurred by accident. In 1876, he was lecturing at the Franklin Institute. In one of his lectures, he was demonstrating low-voltage current from a Leyden jar when two copper wires in loose contact vaporized at the point of contact and were welded together. Thomson observed this and kept it in mind for eight years. Later, when alternating-current generators became available, he perfected the resistance method and apparatus of electric welding, resulting in U.S. Patent No. 347,140. During the life of the patent, the electric welding apparatus was not sold outright but was leased on terms that included a royalty charge for every joint that was welded by the apparatus.

Thomson received many awards while he was alive. He was awarded honorary degrees by Yale University in 1890, by Tufts University in 1894, by the University of Wisconsin in 1904, by Harvard University in 1909, by the University of Pennsylvania in 1924, and by the Victoria University of Manchester, England, in 1924. He was elected to the National Academy of Sciences and served as acting president of the Massachusetts Institute of Technology. He received the John Scott Medal of the City of Philadelphia, the Franklin and Elliott Cresson Medals of the Franklin Institute, the John Fritz Medal of the American Engineering Societies, the Edison Medal of the American Institute of Electrical Engineers, the Lord Kelvin Medal of the English Engineering Societies, and the Faraday Medal of the Institution of Electrical Engineers of London.

Thomson died at his home in Swampscott, near Lynn, Massachusetts, on March 13, 1937, and was buried in Pine Grove Cemetery, Lynn.

Elevator

The New York skyline would look much different today had it not been for the invention of the elevator. The man most responsible for the forerunner to the modern-day elevator was Elisha Graves Otis, who on January 15, 1861, received the first U.S. patent for a safety-type passenger elevator. The number was 31,128.

The idea of vertical travel was not new when Otis built his first elevator. Archimedes and Leonardo da Vinci were associated with this kind of transportation. In the first century B.C., a Roman architect-engineer described a lifting platform with pulleys that operated by human, animal, or water power. In the 17th century, the French architect, Velayer of Paris, built the "flying chair" that operated by a rope running around a wheel at the top of the building. One end of the rope was tied to a chair and the other end to a counterweight. The passenger would throw off a sandbag that was attached to the chair so that the counterweight would descend and the chair and passenger would rise. Robert Dunbar of Buffalo, New York, built a grain elevator operated by steam in 1842. In 1850, Henry Waterman of New York City built a platform-type elevator used to hoist barrels upstairs in the mill of Hecker and Brother. Other Americans associated with early forms of elevators were William Baxter with the electric elevator and Cyrus W. Baldwin with the vertical-geared hydraulic electric elevator. However, the first practical safety-featured passenger elevator was invented by Elisha Graves Otis in 1861.

Elisha Graves Otis was born August 3, 1811, on his father's farm in Halifax, Vermont. He attended common school in Halifax. At age 19, he went to work for a construction company in Troy, New York. After five years, he was forced to give up this job because of ill health. He then went into the trucking business and hauled goods between Troy, New York, and Brattleboro, Vermont. After three years in this business, he bought some land on the Green River in Vermont, where he built a house and sawmill and began to manufacture carriages and wagons.

In 1845, ill health again caused him to change jobs. He moved to Albany, New York, where he worked as a master mechanic in a bedstead factory. Three years later, he opened his own machine shop. It was there that he invented a turbine water wheel. In 1851, Otis moved to Bergen, New Jersey, to become master mechanic in another bedstead factory. The following year, this company opened a new factory

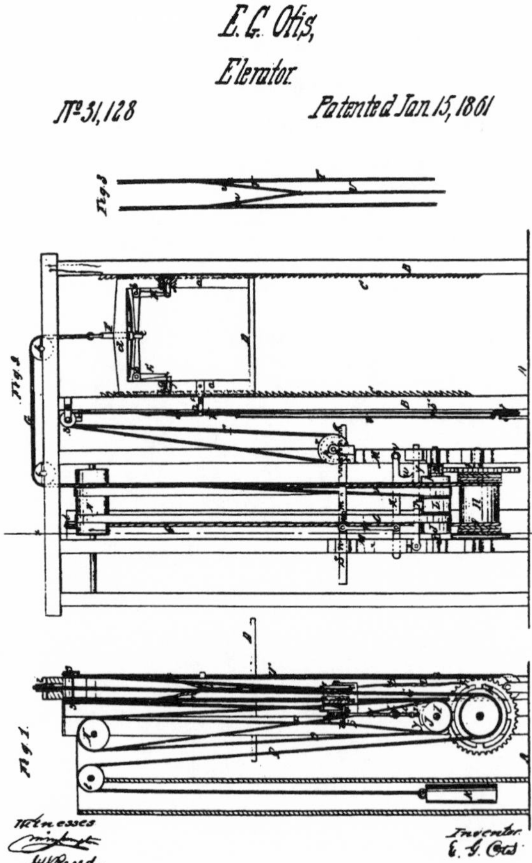

Elisha Graves Otis received the first U.S. patent for a safety-type passenger elevator on January 15, 1861

in Yonkers, New York, and Otis and his son Charles were sent to oversee the erection of the building and the installation of the machinery. It was there, in 1852, that the inventor built his elevator and incorporated a new safety device. This was an ingenious mechanism that consisted of two metal pieces fastened to the elevator platform. These metal pieces remained retracted as long as the cable or rope holding the platform was taut, but if the cable or rope broke, the pieces would spring out and stop the elevator's downward motion. Otis successfully demonstrated his safety elevator to many amazed spectators at the Crystal Palace Exposition in New York in 1853, when, as a passenger, he deliberately cut the rope holding the platform.

Encouraged by his son Charles, Otis gave up his position with the bedstead factory and opened a shop in Yonkers, New York, to manufacture his safety elevator. In 1957, he installed the first safety elevator for passengers in the E. V. Haughwort & Company of New York, a department store. After the premature death of Otis in 1861, another son, Norton, joined with Charles to form Otis Brothers & Company, which later became the Otis Elevator Company. In 1980, the company operated 29 plants worldwide, with over 44,000 employees. It is now a subsidiary of United Technologies.

Otis had other inventions besides the elevator. He received patents for railroad trucks and brakes, a steam plow, and a bake oven. But it was the elevator with its safety feature that brought him fame and his sons fortune.

Ethyl Gasoline

It is well known by many people that the addition of tetraethyl lead to gasoline inhibits the knock of internal combustion engines. What is not so well known is that the man most responsible for this discovery is Thomas Midgley Jr., who received the first U.S. patent for an anti-knocking fuel. The number was 1,573,846 which issued February 23, 1926. This patent was the result of a continuation-in-part application originally filed April 27, 1921.

Thomas Midgley Jr. was born at Beaver Falls, Pennsylvania, on May 18, 1889. His grandfather, James Ezekiel Emerson, was the inventor of the inserted-tooth circular and band saw. In 1896, the family moved to Columbus, Ohio, where Midgley's father, Thomas Midgley, who was also an inventor, designed and manufactured wire wheels and rubber tires in his own factory called the Midgley Tires Company. Thomas Jr. attended the public schools of Columbus and the Betts Academy at Stamford, Connecticut. He graduated from Cornell University with a degree in mechanical engineering in 1911. He worked for a year at the National Cash Register Company in Dayton, Ohio. He then joined his father's company as chief engineer and superintendent. In 1916, Midgley went to work for the Dayton Engineering Laboratories Company (Delco), recently established by Charles F. Kettering, the inventor of the automobile self-starter. His first assignment was to investigate the cause of knock in gasoline engines.

The search for the cause of engine knock consumed many years, involved many people, and required thousands of experiments. The search started in 1912 when Charles Kettering of the Dayton Engineering Laboratories Company, which in 1920 became the General Motors Research Corporation, sought the cause and cure of engine knock. In 1916, a young man came to work for Kettering named Thomas Midgley Jr., a graduate in mechanical engineering from Cornell University. This problem of engine knock was immediately turned over to Midgley and his associate, T. A. Boyd. Working with Midgley and Boyd were Dr. Carroll A. Hochwalt, Dr. James P. Andrew, Charles Harding, and Russell D. Wells.

They soon discovered that the fuel, and not any mechanical defect in the engine, caused the engine knock. Since this was a chemical problem and not a mechanical one, Midgley, with Kettering's approval, employed university professors as consultants in the search for an anti-knock agent as well as answers to other problems associated with engine knock. These

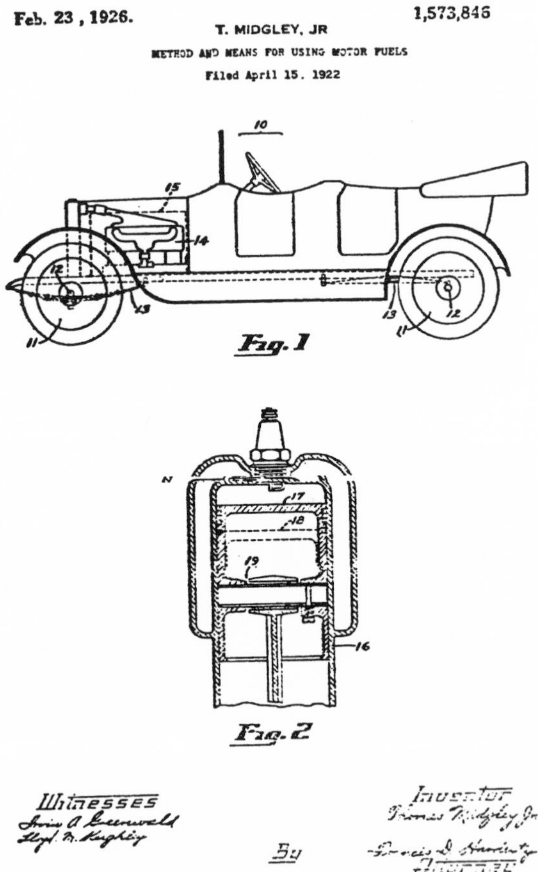

The discovery of anti-knocking fuel is credited to Thomas Midgley Jr., who received this patent on February 23, 1926

included Dr. William E. McPherson, Dr. C. E. Boord, and Henry C. Lord of Ohio State University; Dr. W. C. Ebaugh of Denison University; Dr. Harold Sibbert of Yale University; Professor Victor Lenher of the University of Wisconsin; Dr. Robert E. Wilson, Dr. C. S. Venable, W. G. Whitman, and Robert Haslam of the Massachusetts Institute of Technology; Dr. E. E. Reid of Johns Hopkins University; Dr. Charles Kraus and Dr. C. C. Callis of Clark University; Dr. Wilder D. Bancroft of Cornell University; Dr. Robert A. Kehoe of the University of Cincinnati; and Dr. Graham Edgar of the University of Virginia. These men, in association with Midgley and his team, discovered that tetraethyl lead was the best anti-knock additive for gasoline and that it could be made in sufficient yield to be commercially feasible.

The first test of tetraethyl lead as an anti-knocking agent for gasoline was made in the laboratories of the General Motors Research Corporation on December 9, 1921. However, the problem of undesirable lead deposit on the valves soon developed. This was solved by adding ethylene dibromide to the mixture. The first public sale of Ethyl gasoline, the name recommended by Charles Kettering, was in Dayton, Ohio, on February 2, 1923. In 1924, General Motors and Standard Oil of New Jersey set up a jointly-owned company, called the Ethyl Corporation, to market the product. Billions of gallons of leaded gasoline were sold in the United States until recent environmental laws curtailing the addition of lead to gasoline greatly reduced this volume.

Midgley made other contributions to the world, the major one being the discovery of dichlorodifluoromethane, called Freon, as a refrigerant. He devised a procedure for extracting bromine from the ocean, a process later perfected by the Dow Chemical Company. He also did considerable research on synthetic rubber.

Midgley received many honors. He was elected to the National Academy of Sciences in 1942 and president of the American Chemical Society in 1944. The College of Wooster in 1936 and Ohio State University in 1944 conferred honorary doctorates on him. He received the Nichols Medal in 1923, the Perkin Medal in 1937, the Priestley Medal in 1941, and the Willard Gibbs Medal in 1942.

Midgley suffered an attack of poliomyelitis in the fall of 1940 that left him crippled. He died November 2, 1944, as a result of strangling in the harness of cords and pulleys he had devised to get into and out of bed. He was buried in Greenlawn Cemetery, Columbus, Ohio.

Fertilizer

The first U.S. patent, No. 26,196, for artificial fertilizer was granted to James Jay Mapes on November 22, 1859.

The history of fertilizer goes back to ancient Greece where Xenophon, the historian, recommended to the farmers of ancient Greece that they use green manure to increase crop production. The first white settlers in America found the Indians using guano and fish for fertilizing purposes. In 1665, Sir Kenelm Digby was the first to use chemical fertilizers when he applied saltpeter to his crops. In 1804, Nicholas Theodore de Saussure of Switzerland wrote that the ash ingredients of plants taken from the soil were essential for plant growth. His work was verified by Boussingault on his farm at Alsace. In 1835, Escher of Germany suggested that the fertilizer value of bone could be increased by treating the bone with sulfuric or hydrochloric acid, and in 1839, Baron Justus von Liebig of the University of Giessen in Germany demonstrated that it could be done. Liebig also demonstrated the necessity of supplying plants with phosphorus and potassium. In 1842, Sir John Bennet Lawes received a British patent for the treatment of ground phosphate rock with sulfuric acid and is considered the founder of the chemical fertilizer industry. In 1843, Lawes began the manufacture of super-phosphate at Deptford Creek, London, England. He later sold his fertilizer plants in 1872 for $1.5 million. In 1849, John Pitkin Norton, professor and founder of agricultural chemistry in the Yale Scientific School, published the first American book on agriculture called *Elements of Scientific Agriculture* which had a section dealing with the sources of nitrogen in plants. In 1849, Dr. P. S. Chappell and William Davidson of Baltimore, Maryland, manufactured and sold the first mixed fertilizers in the United States. In 1856, Thomas Green Clemson, an American, wrote that guano or any other manure is good for plants because of the presence of ammonia, which gives off nitrogen. Clemson later became superintendent of agricultural affairs of the U.S. Patent Office in 1860. In 1857, Lawes, Gilbert and Pugh of the Rothamsted Experiment Station, England, demonstrated beyond a doubt the essential nature of ammonia and nitrogen for plant growth. Two years later, on November 22, 1859, James Jay Mapes received the first U.S. patent for artificial fertilizer.

James Jay Mapes was born in Maspeth, Long Island, New York, on May 29, 1806. His father, Jonas Mapes, served as major general in command of

UNITED STATES PATENT OFFICE.

J. J. MAPES, OF NEWARK, NEW JERSEY.

IMPROVEMENT IN FERTILIZERS.

Specification forming part of Letters Patent No. **26,196**, dated November 22, 1859.

To all whom it may concern:

Be it known that I, J. J. MAPES, of Newark, New Jersey, have invented an Improved Fertilizer called the "Improved Superphosphate of Lime," of which the following is a full, clear, and exact description.

By the analysis of soils and plants it is well known what are their constituents, and it is also well known that the continued cropping of soils with less amounts of amendments than those removed as constituents of the crops must gradually impoverish the soil, and in consequence we find the uneven or unequal removal of these constituents to render the soil unfit for the raising of wheat and other crops requiring the presence of phosphoric acid, lime, potash, &c., they being among the first of the natural constituents of the soil parted with, and hence the continued reduction of the wheat and other crops per acre in the older States has occurred.

The constituents of the fertilizer invented by me are such as more nearly to supply these deficiencies than any other before known, and under my combination they are in a less volatile state, so as to admit of a more economic use. The supply of ammonia, so necessary as a stimulant to cause plants to appropriate their inorganic wants when within their reach, in a soluble and non-volatile form, and properly combined with the inorganic necessities of the soil and plants, is new and original.

The nature of my invention consists in saturating phosphate of lime—such as mineral apatite or calcined bones—with sulphuric acid to such extent as to form the superphosphate of lime by combining part of the lime with the sulphuric acid as sulphate of lime, leaving the whole of the phosphoric acid combined with the remaining portion of lime as superphosphate of lime, when such products are combined with guano and sulphate of ammonia, or the equivalents thereof, such as the ammoniacal liquor of gas-works or other ammoniacal waste.

I take one hundred parts, by weight, of apatite or calcined bones or phosphate of lime, in a suitable vessel and saturate it with sulphuric acid, which will require about fifty-six parts, by weight, of the ordinary sulphuric acid of commerce, and after the superphosphate of lime is formed by part of the lime combining with the sulphuric acid as sulphate of lime, and the phosphoric acid combining with the remaining portion of lime as superphosphate of lime, I then add and thoroughly mix therewith thirty-six parts, by weight, of Peruvian guano and twenty parts, by weight, of sulphate of ammonia; or the ammonia may be added by equivalent additions of the ammoniacal liquor of gas-works or other ammoniacal waste.

It is found that the amount of sulphuric acid here prescribed for converting the bone earth into superphosphate affords sufficient excess of acid to convert the carbonate and other unstable salts of ammonia into the sulphate of ammonia, and thus fix these volatile ingredients of the Peruvian guanos and secure their fertilizing qualities to the soil and crops.

This compound thus produced constitutes a fertilizer more nearly corresponding with the general deficiencies of soils than any other heretofore known, and therefore better adapted to furnish the ingredients required, while at the same time the volatile carbonates of the guano are changed into the non-volatile sulphates, and thus detained in the soil until used by the plants, instead of being dissipated into the atmosphere, as when used in the ordinary methods heretofore practiced.

I do not design to limit myself to the exact proportions of the ingredients herein named, as the chemical constituents may vary in their composition. The crude products may be more or less pure and require more or less of the other ingredients to neutralize them.

Having described the nature of the invention and shown the several peculiarities of the process for fixing the ammonia and making a concentrated manure, I design to cover these materials or their equivalents.

I do not claim the use of sulphuric acid for preparing the superphosphate of lime, nor yet the use of sulphate of ammonia as a manure, nor the mixture of one guano with another, as such use and such mixtures are known; but

What I claim as my invention, and desire to secure by Letters Patent, is—

The production of a fertilizer by combining guano and sulphate of ammonia or its equivalent with burned bones, or their equivalent when the said bones or equivalent have been treated by sulphuric acid, as specified, the whole being prepared substantially in the manner and for the purpose set forth.

JAS. J. MAPES

Witnesses:

WALTER B. NORTH,
SAML. GRUBB.

James Jay Mapes obtained the first U.S. patent for artificial fertilizer on November 22, 1859

all defense troops in the New York area during the War of 1812. After the war, General Mapes was a senior partner in a New York firm of importer and merchant tailors, a founder and director of the Bank of Savings of New York, and was associated with DeWitt Clinton in the Erie Canal project.

At the age of eight, young James made illuminating gas in a retort improvised out of a piece of clay pipe after hearing a lecture on the subject. When he was 11 years old, he was sent to Hemstead, Long Island, to board with the family of William Cobbett to attend the classical school of Dr. Timothy Clowes. In his teens, he entered his father's business as a clerk. In 1832, he invented several improvements in the methods of refining sugar. In 1834, he quit the mercantile business to open an office as a consultant in analytical chemistry, and was frequently asked for expert testimony in chemical patent litigations. The New York State Senate commissioned him to make a series of analyses of beer and wine. While serving as a consultant, he made several important contributions to a wide variety of chemical industries, including experiments with Seth Boyden that led to Boyden's invention of malleable iron.

Between 1835 and 1838, in addition to his consulting work, Mapes was professor of chemistry and natural philosophy at the National Academy of Design in New York. From 1840 to 1842, he edited the *American Repertory of Arts, Sciences, and Manufactures* in four volumes. In 1843, he became associate editor of the *Journal of the Franklin Institute* of Philadelphia. In 1844, he was elected president of the Mechanics Institute of the City of New York. He was instrumental in the founding of night schools for adult education, the forerunners of such schools as Cooper Institute.

Mapes turned his attention toward scientific agriculture and advocated sub-soil drainage, crop rotation, and the use of chemical fertilizers. In 1847, he purchased a worn-out farm in the Weequahic section of Newark, New Jersey, and set to work to prove his theory by actual practice. In 1849, he founded and edited until 1863 a journal called *The Working Farmer,* in which he published the results of his experimental farming. He organized state and county agricultural societies and was the first to advocate a federal Department of Agriculture with a secretary who should be a cabinet officer. He transformed the poor, worn-out soil of his Newark farm into highly productive acres, growing both vegetables and field crops under carefully controlled conditions. His farm became the model of the modern Agricultural Experiment Stations. He took pupils in small groups of three or four for a month's "short course" in scientific farming. He invented a sub-soil plow and developed a formula for nitrogenized superphosphate which was the first complete plant food artificial fertilizer used in the United States. As early as 1849, he produced this superphosphate for use on his own farm but, due to considerable litigation, did not receive a patent until November 22, 1859. Mapes never tried to enforce this patent, but encouraged others to make and market similar fertilizer mixtures. His son, Charles V. Mapes, later manufactured and sold his fertilizer.

James Jay Mapes died in New York City on January 10, 1866. His friend, Horace Greeley, writing an editorial on his death, said of him, "American agriculture owes as much to him as to any man who lives or has ever lived." He was survived by his wife, three daughters, and one son. One of his daughters, Mary Mapes Dodge, was editor of *St. Nicholas Magazine* and the author of the children's classic, *Hans Brinker*. His son, Charles Victor Mapes, succeeded him as a fertilizer manufacturer and agricultural chemist. Among his grandchildren are a distinguished physician, James Jay Mapes II, the first to use diphtheria antitoxin in America; Victor Mapes, the playwright; and Clive Mapes, a pioneer in the radio industry.

First Patent in America

Long before there was a United States Patent Office, or even a United States, the colonies granted patents to certain individuals. The first patent granted in America for a process invention was given to Samuel Winslow in 1641 by the Massachusetts Bay Colony for a process of manufacturing salt. The first patent granted in America for a mechanical invention was given to Joseph Jenks in 1646 by this same colony for improved sawmills and scythes.

In 1641, the Massachusetts Bay Colony General Court passed a system of laws called the *Body of Liberties*. One section of these laws decreed: "There shall be no monopolies granted or allowed among us, but of such new inventions as are profitable to the country, and that for a short time." This formed the basis for granting the first patents in America.

On June 2, 1641, this same General Court granted a patent to Samuel Winslow for a process of manufacturing salt. The Court decreed: *"Whereas Samu; Winslow hath made a proposition to this Court to furnish the countrey with salt at more easy rates then otherwise can bee had, & to make it by a meanes & way weh hitherto hath not bene discovered, it is therefore ordered, that if said Samu; shall, wthin the space of one yeare, set upon the said worke, hee shall enjoy the same, to him & his assosiats, for the space of 10 yeares, so as it shall not bee lawfull to any other pson to make salt after the same way during the said yeares; pvided, nevrthelesse, that it shall bee lawfull for any pson to bring in any salt, or to make salt after any othr way, dureing the said tearme."*

In 1646, the General Court again granted a patent to Joseph Jenks for improved sawmills and scythes. The Court resolved that: *"In answer to the petition of Joseph Jenckes, for liberty to make experience of his abillityes & inventions for ye making of engines for mills, to goe wth water, for ye more speedy dispatch of worke than formerly, & mills for ye making of sithes & other edged tooles, wth a new invented sawmill, that things may be afforded cheaper than formerly, & that for fowerteene yeers without disturbance by any others setting vp the like inventions, that so his study & costs may not be in vayne or lost; this petition was graunted, so as power is still left to restrayne ye exportacon of such manufactures, & to moderate ye prizes thereof if occacon so require."*

Other colonies, following the lead of Massachusetts, also granted patents to their inhabitants. In 1652, Virginia granted to George Fletcher the first patent from that colony for a method of distilling and brewing with wooden vessels. South Carolina granted the first patent to Peter Jacob Guerard in 1691 for a pendulum engine to husk rice. This was the first

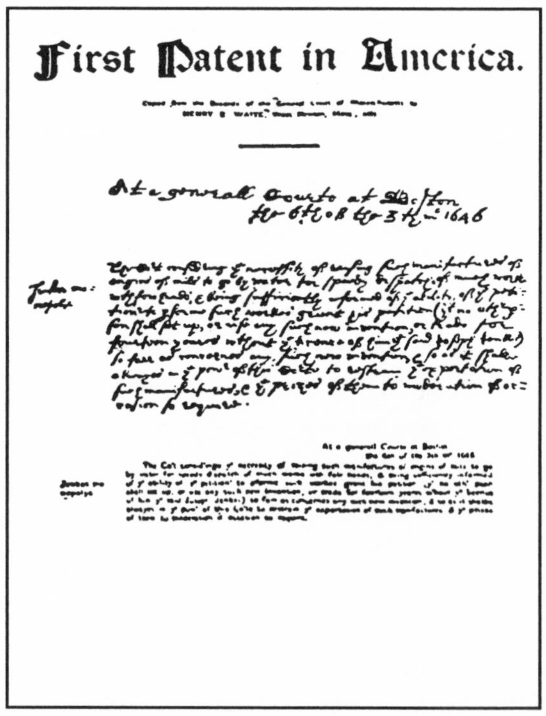

First Patent in America.

The first patent for a process invention was given to Samuel Winslow in 1641 by the Massachusetts Bay Colony for a process of manufacturing salt.

patent in America to make provision for licensing and to specify a procedure for legal remedy against infringement. In 1712, New York granted the first patent to John Parmiter for the manufacture of lampblack. The colony of Connecticut granted its first patent for an invention to Ebenezer Fitch in 1716 for a slitting mill. In 1731, Rhode Island granted the first patent to John Lucena, who was from Portugal but was naturalized the same year, for the manufacture of soap. The Maryland colony in 1770 granted to John Clayton and to Isaac Perkins the first patents from that colony for threshers.

The colonies of New Hampshire, Georgia, Delaware, New Jersey, Pennsylvania, and North Carolina granted no patents before the Revolutionary War. However, all these colonies or states, except North Carolina, granted patents before the first U.S. patent laws were formulated in 1790. New Hampshire, in 1786, granted a patent to Robert Dearborn for a printing press. In 1788, Georgia granted to Isaac Briggs and William Longstreet a patent for improvement in steam engines. Both Delaware and New Jersey granted, in 1788, to John Fitch a patent for improvement in steamboats. In 1789, Pennsylvania granted a patent for improvement in steam engines to James Rumsey.

History is silent about the life of Samuel Winslow except that he was living in the Massachusetts Bay Colony in 1641. In response to a need, Winslow discovered a more efficient and cheaper way of supplying salt to the colonists. He petitioned the Massachusetts General Court for a patent, contending that his process was new. On June 2, 1641, the General Court granted him a patent for a term of 10 years. This grant stated that during its 10-year term all other salt-making processes were to remain open to public use, but it was not lawful for others to make salt by the Winslow process.

Joseph Jenks was born in Colnbrook, England, in 1602. Not much is known of Jenks until 1642, when he was employed at Hammersmith, England, as an iron-worker. In 1643, he was induced by John Winthrop Jr., whose father was the Governor of the Massachusetts Bay Colony, to come to America to assist in the development of the first iron works to be established in America. He built an iron foundry and forge at Saugus, near Lynn, Massachusetts, on the lands of Thomas Hudson. He was the first builder of machinery in this country and the first worker in brass and iron on the continent. His first casting was an iron quart pot made for the Hudson family. In 1652, he was commissioned to cast a die at his ironworks for the first money coined in America. The dies were cast for three-pence, six-pence, and a shilling. This famous pine-tree shilling, reportedly designed by his daughter-in-law, showed the word "Masatusets" and a pine tree on one side and "New England, Anno 1652" stamped on the reverse side. In 1654, he built for the town of Boston the first water-carrying fire engine used in America. In 1655, he was granted his second patent for a grass scythe. The old English scythe was short and thick. Jenks made the blade thinner and longer with the back thickened by a welded bar of iron. There has been practically no change on the shape or size of the scythe to this day.

Joseph Jenks died at Saugus, Massachusetts, in March of 1683. His son, Joseph Jenks II, founded the town of Pawtucket, Rhode Island, and became a member of the governor's council and served in the house of deputies for several years. His grandson, Joseph Jenks III, served as Governor of Rhode Island from 1727 to 1732.

FM Radio

The opportunity to listen to static-free, high-fidelity music on the radio is due to the work of Edwin Howard Armstrong who received the first U.S. patent, No. 1,941,066, on December 26, 1933, for FM radio circuitry.

The discovery of radio waves came about as a result of the study of light. In 1865, James Clerk-Maxwell, a Scottish physicist, mathematically described light as a series of waves traveling through space. In 1887, Heinrich Hertz, a German physicist, discovered that a rapidly vibrating electric charge will produce electromagnetic waves that carry both electric and magnetic energy and that these waves travel at the speed of light. He built a simple transmitter consisting of a primary coil that had a few turns of thick copper wire and a secondary coil that had many turns of fine copper wire. He connected the primary coil to a battery to establish a steady current flow. Then, to form a spark gap, he connected the secondary coil to a pair of metal balls separated by a few inches. A metal plate attached to each ball acted as an antenna. As the flow of current in the primary coil was interrupted, a very high voltage was induced in the secondary coil by electromagnetic induction, causing a spark between the balls. This created very high frequency oscillating currents to radiate from the metal plates. Thus was born electromagnetic radiations, now called radio waves, that traveled through space. Hertz built a simple receiver to verify that his transmitter was indeed sending electromagnetic waves through space.

In 1895, Guglielmo Marconi, an Italian, recognized that long-distance communication would be possible if these hertzian waves were properly utilized. Previous to 1895, David Hughes and Alexander Lodge in England and Alexander Popov in Russia all built wireless telegraphs. However, it remained for Marconi to usher in the age of radio. He improved the Hertz receiver circuitry and developed an extremely sensitive metal-filings coherer, invented by Edouard Branly of France in 1891. This was a long tube filled with iron filings which had increased electrical resistance when struck by radio waves. He also built a mechanical interrupter to repeatedly break the flow of current to the primary coil to generate dots and dashes. Marconi's first wireless patent was issued in 1896 in Italy. The first messages traveled only a few hundred feet, but in 1898 the distance had increased to 32 miles. By 1899, a message was sent and

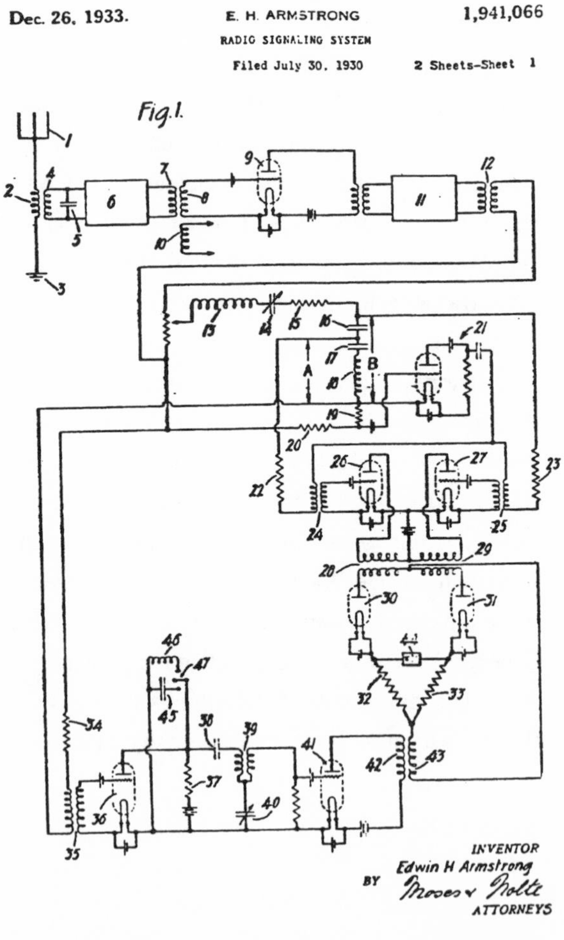

Dec. 26, 1933. E. H. ARMSTRONG 1,941,066
RADIO SIGNALING SYSTEM
Filed July 30, 1930 2 Sheets-Sheet 1

Fig. 1.

INVENTOR
Edwin H Armstrong
BY Moses & Nolte
ATTORNEYS

The patent for FM radio circuitry was granted to Edwin Howard Armstrong on December 26, 1933

received across the English Channel. At 12:30 p.m. on December 12, 1901, his associate, G. S. Kemp, sent a message from the coast of England, and Marconi picked it up on his receiver atop Signal Hill in St. John's, Newfoundland.

Sparks and metal-filings made possible the sending and receiving of dots and dashes, but voice communication required much more sensitive equipment. This was made possible by the invention of crystal detectors by G. W. Pickard, an American, in 1902; vacuum tubes by J. A. Fleming of England in 1904; the generation of a clean, sine-wave, single high frequency, RF oscillation to act as a carrier for the human voice by Reginald Aubrey Fessenden, a Canadian, in 1906; and the amplifying or triode vacuum tube by Lee de Forest, an American, in 1907. Then came Edwin H. Armstrong, an American, on the radio scene. He invented the regenerative or feedback circuit in 1912 and the heterodyne circuit in 1918. But perhaps his greatest invention was in 1933 when he invented the FM circuit.

Edwin Howard Armstrong was born December 18, 1890, in the old Chelsea district of New York City. His father, John Armstrong, worked for the Oxford University Press. In 1902, the family moved to 1032 Warburton Avenue, Yonkers, New York. In 1904, Armstrong's father gave him two books that greatly influenced his life. The books were *The Boy's Book of Invention* and *Stories of Inventors.* At the age of 14, he determined to become an inventor but could not decide between wireless telegraphy or X-rays until entering Yonkers High School in 1905. He chose wireless, set up a wireless amateur station in the attic of his Yonker's home, and became widely known in amateur radio as early as 1906.

He entered Columbia University in 1909 as a student in electrical engineering. The teacher who had the most influence on him was Professor Michael Idvorsky Pupin, who not only was the founder of Columbia's department of electrical engineering but also the inventor of the Pupin loading coil. This was a device for stepping up the electromotive force of a current, which Pupin sold to the American Telephone & Telegraph Company. Pupin encouraged Armstrong's activities in wireless and became a lifelong friend. Throughout his student days at Columbia, young Armstrong never ceased to experiment at night and on weekends in his attic room, trying to discover a method of increasing the signal of the wireless. He graduated from Columbia University in 1913 with a degree in electrical engineering but remained at Columbia for a period of time on a fellowship.

On September 22, 1912, he began a series of experiments that would lead to the first of four pioneering inventions. In the winter of 1913, he told his friend and classmate, Herman Burgi, that he had discovered some new circuits that would greatly amplify wireless signals. He was advised to seek a patent. However, the cost of filing for a patent at that time was $150. After a canvas of friends, relatives, and resources, he found that he could not raise the money. His uncle, Frank Smith, suggested that he prepare a sketch of his device and have it witnessed and notarized. Armstrong drew up a diagram of his circuit and, with Herman Burgi as witness, had it notarized on January 13, 1913.

Late in 1913, after finding the money, he filed for his first patent. Within a short time, two other inventors, Irving Langmuir of General Electric and Alexander Meissner of Germany, filed applications for the same or similar inventions. The U.S. Patent Office instituted interference proceedings to determine the first inventor. Armstrong was able to establish priority over these two through the notarized and witnessed drawing suggested by his uncle. On October 6, 1914, he was issued U.S. Patent No. 1,113,149 for the regenerative or feedback receiver circuit, a circuit still used in modern radio receivers.

Armstrong discovered that if part of the plate's output current in a vacuum tube was fed back and tuned into the grid, it strengthened the incoming signals to the grid as much as a thousand times. For this, he received a patent on the regenerative or feedback receiver circuit. He also discovered that when the feedback was increased beyond the point of maximum amplification, the tube suddenly changed from a receiver to a transmitter by rapidly oscillating the current from the filament to the plate to send out electromagnetic waves of its own. He filed a patent application on this concept at the same time that he had filed on the feedback receiver. However, Lee de Forest filed applications about the same time for the same invention. These applications became involved in interference proceedings in the Patent Office, and for the next several years, a messy litigation existed between Armstrong and de Forest on who owned the patent on the regenerative circuit as a transmitter. Finally, the Supreme Court in 1934 handed down a decision against Armstrong.

During World War I, Armstrong was a major in the U.S. Army Signal Corps. While in France, he discovered that when two high-frequency waves interact, they produce a signal wave of intermediate frequency equal to the difference in frequency of the two mixed waves. By amplifying this intermediate frequency, additional receiver sensitivity and signal strength is achieved. He developed this superheterodyne receiver for military purposes to be used to detect incoming planes by the detection of the electromagnetic waves produced by their ignition system. At this time, the U.S. Army allowed inventions made by inventors in the army to be retained by the inventors. On February 8, 1919, Armstrong filed an application on the superheterodyne. It was issued on June 8, 1920, as U.S. Patent No. 1,342,885.

After World War I, he returned to Columbia University to work with Pupin on the problem of static elimination. He soon found himself in debt to the tune of $40,000 for lawyer's fees over the de Forest litigation. On October 5, 1920, Armstrong sold his feedback and superheterodyne patents to Westinghouse for $335,000. This money enabled him to pay off his debts and served as a cushion for further research.

In 1921, he discovered the superregenerative circuit and was issued U.S. Patent No. 1,424,065 for this invention. In 1922, Sarnoff of the Radio Corporation of America offered Armstrong $200,000 in cash and 60,000 shares of R.C.A. stock for his superregenerative circuit patent. He promptly accepted and during the next 20 years he gained some $9 million from this offer.

In 1933, Armstrong achieved perhaps his greatest invention. For 20 years, he had searched for a way to eliminate static. Up to 1933, all radio signals were superimposed on a station's carrier wave by varying or modulating the carrier wave's power or amplitude into an exact replica of the sound waves being transmitted. This is known as amplitude modulation or AM. Thunderstorms and electrical appliances will also randomly modulate the amplitude of the carrier wave of the radio signal, resulting in static. Armstrong devised an entirely different radio system. He devised a method, called frequency modulation or FM, in which the transmission of a radio signal is done by modulating the frequency of the carrier wave according to the sound waves coming from the microphone. The incoming signal is first heterodyned and amplified. This signal is then passed to a limiter that clips off any amplitude static and sends this signal to the discriminator. The discriminator then converts the frequency variation into amplitude variations for detection and amplification into sound at the loudspeaker.

On December 26, 1933, four U.S. patents were issued to Armstrong on the same day for this invention, having consecutive numbers from 1,941,066. Ordinarily, industry would have welcomed such a system, but with millions of dollars already invested in AM broadcasting, the few industrial giants that controlled AM broadcasting were not ready to welcome a new radio broadcasting system. They feared it would compete or conceivably replace their system. Armstrong faced a long, hard battle in the courts in the years ahead by defending the validity of his patents and filing infringement suits against companies that used his patents without paying royalties. Although he or his heirs eventually prevailed, the lawsuits and the thought that there was a conspiracy against him took their toll. On February 1, 1954, in a fit of depression or despair, after writing a letter to his wife, Armstrong jumped to his death from his apartment window, 13 stories above the street. He was buried in Locust Grove Cemetery at Merrimac, Massachusetts.

Many honors came to Armstrong. In 1917, the Institute of Radio Engineers awarded him the Medal of Honor; in 1929, Columbia University gave him an honorary doctorate; in 1939, Columbia University bestowed on him the Egleston Medal; in 1940, the American Society of Mechanical Engineers awarded him the Holley Medal; in 1941, Muhlenberg College conferred on him an honorary doctorate; in 1941, the Franklin Institute gave him the Franklin Medal; in 1941, the City of Philadelphia awarded him the John Scott Medal; and in 1943, the American Institute of Electrical Engineers gave him the Edison Medal. He was also the first recipient of the Armstrong Medal in 1935. The Union Internationale des Telecommunications in Geneva, Switzerland, in May 1955, added the name of Edwin Howard Armstrong to the list of great men in electricity and communications. His name appears in the company of Alexander Graham Bell, Samuel F. B. Morse, Michael Pupin, Nikola Telsa, Andre Marie Ampere, Michael Faraday, Karl Friedrich Gauss, Heinrich Hertz, Lord Kelvin, Guglielmo Marconi, and James Clerk-Maxwell.

Frozen Food

When shoppers purchase a package of Birdseye frozen food, most think that it is the name of the product and not that of the inventor, Clarence Birdseye, who received the first U.S. patent for this product. The number was 1,773,080, issued August 12, 1930.

Food preservation has drawn the attention of people since the beginning of time. Such methods as salting, pickling, smoking, and curing have been used. Home canning, commercial canning, the family cellar, cold spring water, the mobile refrigerator, the icebox, the home refrigerator, and the home deep-freeze have all contributed at one time or another to the preservation of food. However, one of the most successful methods for food preservation was invented, developed, and commercialized by Clarence Birdseye. This method involved quick-freezing meat, seafood, vegetables, and fruit in convenient packages, without altering the original taste. Birdseye's name became a household word and his process created a multibillion-dollar industry.

Clarence Birdseye was born December 9, 1886, in Brooklyn, New York. His father, Clarence Frank Birdseye, was a lawyer and legal scholar. The younger Birdseye attended school in Brooklyn and developed an early interest in natural history. Before he reached his teens, he considered himself an expert in taxidermy. This led him to place an advertisement in a sports magazine announcing the opening of The American School of Taxidermy, where he would teach classes in this art. He attended high school in Montclair, New Jersey, and enrolled in the school's cooking class, an interest that would remain with him the rest of his life.

In 1908, Birdseye entered Amherst College. To meet expenses, he sold frogs to the Bronx Zoo to feed the snakes and live black rats to Columbia University for breeding experiments. During the summers and part of the school year, he worked as a field naturalist for the Biological Survey of the United States Department of Agriculture. His time was spent doing research on Rocky Mountain spotted fever in Montana and trapping wolves in northern Michigan. In 1912, he published a short monograph entitled, *Some Common Mammals of Western Montana in Relation to Agriculture and Spotted Fever*.

That same year, Birdseye went to Labrador to work on the hospital ship of Sir Wilfred T. Grenfell. He remained in Labrador for the next three years, engaging mainly in fur trading and trapping. Then, in

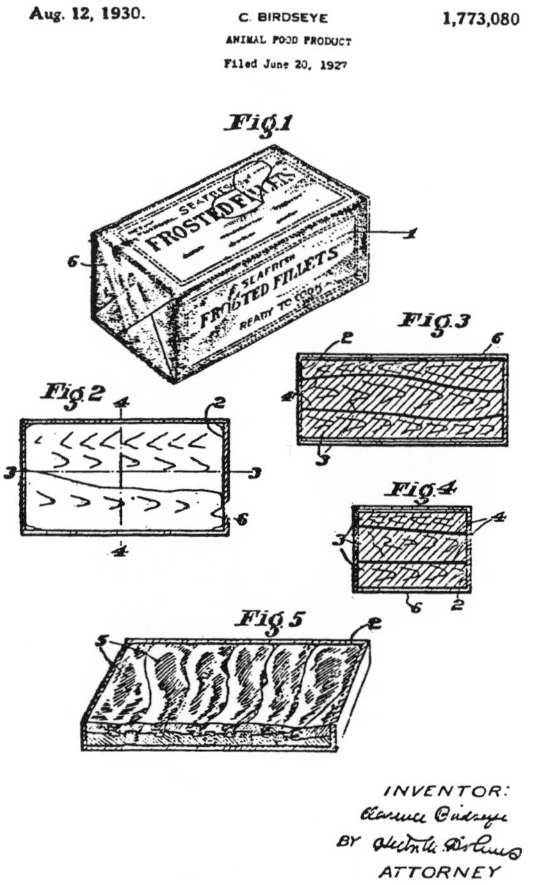

Aug. 12, 1930. C. BIRDSEYE 1,773,080
ANIMAL FOOD PRODUCT
Filed June 20, 1927

Fig. 1 Fig. 2 Fig. 3 Fig. 4 Fig. 5

INVENTOR:
BY
ATTORNEY

Clarence Birdseye became a household name through his invention of a process for quick-freezing, patented August 12, 1930

1915, he returned to the United States to marry Eleanor Gannett of Washington, D.C. One year later, he returned to Labrador with his wife and their five-week-old baby. Wishing that his family could have fresh food all year long, Birdseye turned his attention to food preservation. He discovered that if fresh cabbage and the meat of rabbits, ducks, and caribou were placed in barrels of sea water and quickly frozen by the arctic temperatures of 40 to 50 degrees below zero, they would retain their freshness when thawed and cooked months later. He concluded that it was the rapid freezing in the extremely low temperatures that made the food remain tender and tasty.

In 1917, Birdseye and his family returned to the United States with the idea of continuing his work on the fast-freezing of food. However, World War I interrupted his research and he became purchasing agent for the United States Housing Corporation until 1920. At that time, he became assistant to the president of the United States Fisheries Association.

In 1923, a friend permitted Birdseye to use the corner of his icehouse in New Jersey, where the inventor resumed his quick-freezing experiments with an investment of only seven dollars for an electric fan, buckets of brine, and cakes of ice. The first successful product was dressed fillets of haddock, frozen in square containers made from old candy boxes. In 1924, Birdseye borrowed on his life insurance, interested a small group of investors, and formed the General Seafoods Company in a two-story building on Fort Wharf in Gloucester, Massachusetts. The first floor of the building was devoted to research, and it was here that he developed his patented belt freezer method. By this method, perishable packaged foods of all kinds could be quick-frozen by pressing them between refrigerated metal plates.

In 1929, the Postum Company and the Goldman-Sachs Trading Corporation acquired Birdseye's patents and trademarks for $22 million. Later, the Postum Company bought the Goldman-Sachs Trading Corporation interest and changed the name to the General Foods Corporation. Under the new tradename of Birds Eye Frosted Foods, quick-frozen vegetables, fruits, seafoods, and meat were sold to the public for the first time in 1930 in Springfield, Massachusetts. However, public acceptance was slow. It took the extensive marketing organization of General Foods to overcome the public's widespread aversion to "frozen" foods. Today, Birdseye's name is a famous trademark and quick-frozen foods have become a multibillion-dollar industry.

After disposing of his quick-frozen food operation, Birdseye turned his attention to other interests. He invented an infrared heat lamp and a spot light for store window displays and established the Birdseye Electric Company to market these products. He also invented a harpoon for marking whales. He organized Processes Incorporated, for continuing research and development in food preservation. He received several patents for the process of quick-drying foods and founded Dehydration Incorporated. In 1951, he was co-author with his wife of *Growing Woodland Plants*. In 1953, he went to Peru on an assignment by a paper company to develop a new method for making paper stock from crushed sugar cane stalks. While in Peru, he suffered a heart attack that forced him to return to New York City. He died at his home in Gramercy Square on October 7, 1956.

Glass Bottles

Many items, such as food, beverages, medicines, chemicals, and cosmetics are packaged in glass bottles. This was made possible by Michael Joseph Owens who received the first U.S. patent for an automatic glass bottle making machine. The patent was granted August 2, 1904, and bears the number 766,768.

The first containers were bottles made from animal skins but in about 1500 B.C. glass replaced skin for bottle material in Egypt and Mesopotamia. In about 30 B.C., glass bottles were made along the eastern Mediterranean coast by blowing molten glass at the end of a hollow iron pipe. The glass industry flourished in Persia, Iraq, and Venice around 1200. Bottles made by hand-blowing came to England about this time, and in 1226 the first glass manufacture was done by Laurence Vitraerius in Chiddingfold, Surrey.

The first glass factory in America started in 1608 when the second ship to arrive at Jamestown, Virginia, brought eight skilled glass workers who, under Captain John Smith, built four furnaces to make vials, bottles, drinking glasses, and glass beads for trading with the Indians. Up to about 1890, practically all bottles manufactured in the United States were hand-blown. Even the bulbs for Edison's electric light were blown by Fred Douerlein of the Corning Glass Works in New York. In 1887, Ashley, an English engineer, introduced a semi-automatic glass-blowing machine for making bottles at Castleford in Yorkshire. In 1904, Michael J. Owens revolutionized the bottle-making industry by patenting an automatic machine for making bottles that operated by a combination of vacuum and blowing. This machine could produce 2,500 bottles an hour.

Michael Joseph Owens was born in Mason County, West Virginia, on January 1, 1859. His father was a coal miner, having emigrated to this country some 20 years before. At the age of 10, young Michael was sent to work in the glass factory of Hobbs, Brochunier & Company in Wheeling, West Virginia. By the time he was 15 he had become an expert glassblower.

In 1888, he joined forces with Edward Drummond Libbey in his glass factory in Toledo, Ohio, and soon became manager of a branch factory in Findlay, Ohio. In 1891, he invented a machine that automatically opened and closed the molds for mold-blown glass. He also invented an automatic machine for blowing glass tumblers, electric light bulbs, and lamp chimneys. In 1893, he was placed in charge of the Libbey Glass Company exhibit at the World's

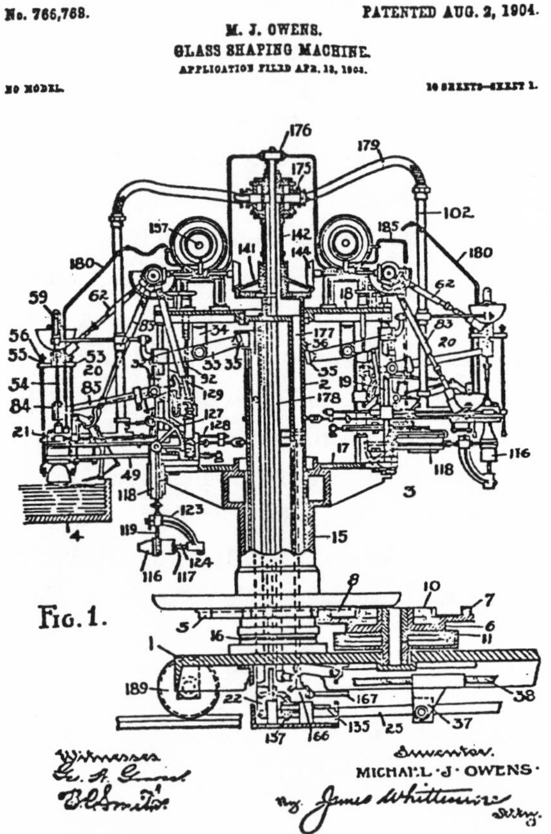

Michael Joseph Owens's Glass Shaping Machine, patented August 2, 1904

Columbian Exposition in Chicago.

About this same time, Owens began a series of experiments which eventually led to a completely automatic bottle-making machine. Improvement was long overdue in this field. Individual bottles were still being made almost exactly as they had been made in ancient Rome. He built a machine based upon the principle of gathering a small amount of molten glass by vacuum and then placing this gathered glass over a desired mold, and by using compressed air, forcing the glass to assume the shape of the mold. He continued to improve this machine so that by 1903 sufficient progress had been made for Owens and Libbey to form the Owens Bottle Machine Company, later called the Owens Bottle Company. In 1904, Owens perfected his automatic bottle-making machine and, until 1917, his was the only fully automatic machine for making all kinds of bottles. It was recognized as the major advance in the art of glass manufacture since the blowpipe was first used 2,000 years before. His machine could produce thousands of bottles a day ranging in capacity from one-tenth of an ounce to 13 gallons.

In 1912, Owens and Libbey purchased the patents of Irving W. Colburn for a machine for the continuous drawing of flat sheet glass. In 1916, they formed the Libbey-Owens Sheet Glass Company and built a factory at Charleston, West Virginia. In 1930, this company merged with the Edward Ford Plate Glass Company to become the Libbey-Owens-Ford Glass Company.

Owens did much for the glass industry, but the glass industry also did much for him. He was vice president of the Owens Bottle Company, the Owens European Bottle Company, and the Libbey-Owens Sheet Glass Company. He was granted 45 United States Patents on apparatus involving the various aspects of glass manufacture.

Michael Joseph Owens died December 27, 1923, in Toledo, Ohio.

Guncotton

An explosive made from cellulose, called guncotton, was the result of work done by Christian Friedrich Schönbein. For this work, he received the first U.S. patent for guncotton as No. 4,874, issued December 5, 1846.

Explosives were not new in 1846. Berthold Schwartz of Germany is credited in 1320 with the invention of gunpowder, although this material was used by the Chinese long before this. In 1832, Henry Braccanot, professor of natural philosophy at the Lyceum in Nancy, France, was the first to do any recorded work on the action of concentrated nitric acid on starch cotton and wood fibers to obtain a solid substance he named xyloidine. Not knowing what he had, he made no further study on this substance. In 1838, Theophile Jules Pelouze of the University of Paris published the results of some work on the effects of nitric acid on starch cotton or wood. This combination produced a substance of highly combustible character that burned quickly and left no residue. Because of preoccupation with other work, however, Pelouze did no further research. It remained for Schönbein, a German professor of chemistry at the University of Basel in Switzerland, assisted by co-worker F. Bottger, to discover the process of nitrating cellulose in the presence of sulfuric acid and to suggest its use as a substitute for gunpowder.

Christian Friedrich Schönbein was born October 18, 1799, at Metzingen, Wurttemberg. The finances of the family of eight children would not permit a lengthy period of schooling for young Christian. He was apprenticed in 1813 to a chemical and pharmaceutical factory at Boblingen. There, he acquired a knowledge of theoretical and applied chemistry as well as Latin, French, English, and mathematics. In 1820, he accepted a position in a chemical factory near Erlanger. At this time, Schönbein's employer, J. N. Adam, made him private tutor to his children. This gave Schönbein time and funds to enter the University of Erlanger in 1821. After one semester, he enrolled at the University of Tubingen but returned to the University of Erlanger in 1823. For three years, he was a teacher of chemistry, physics, and mineralogy at an educational community founded by Friedrich Froebel at Keilhau, Thuringia. In January 1826, he took a position in a boys school at Epsom, England. After a short stay at Epsom, he received an offer to act as a substitute professor of chemistry and physics at the University of Basel. The university conferred a doctorate on him in 1829 and a full professorship of chemistry and physics in 1835. He remained there for 40 years and issued more than 350 scientific publications.

One of Schönbein's most important scientific

UNITED STATES PATENT OFFICE.

O. F. SCHÖNBEIN, OF BASLE, SWITZERLAND, ASSIGNOR TO WM. H. ROBERTSON.

IMPROVEMENT IN PREPARATION OF COTTON-WOOL AND OTHER SUBSTANCES AS SUBSTITUTES FOR GUNPOWDER.

Specification forming part of Letters Patent No. 4,874, dated December 5, 1846.

To all whom it may concern:

Be it known that I, CHRISTIAN FREDERICK SCHÖNBEIN, the undersigned, of Basle, in Switzerland, have discovered a new Process for Preparing Cotton and other Vegetable Fibrous Substances as a Substitute for Gunpowder, of which the following is a full and exact description, reserving to myself the right of adding thereto such other materials of an explosive nature as I may hereinafter add.

What I claim as my discovery and ask a patent for is—

Treating vegetable fibrous substances and other organic matters being of a chemical composition analogous to that of the said vegetable fibrous substances with a mixture of nitric acid of 1.5 specific gravity, or thereabout, and sulphuric acid of 1.85 specific gravity, or thereabout, at the common temperature, or a lower one, by which compounds are formed which on being heated more or less ignite suddenly and produce gaseous matters. Of all the vegetable substances known to me, cotton-wool is the most fit material for producing an explosive compound answering the purposes of gunpowder, and I therefore claim the use of cot-[illegible] purpose when prepared as herein directed.

To produce that explosive compound the following process is to be gone through.

First. Clean cotton is immersed in a mixture containing the nitric and sulphuric acids of the specific gravities before mentioned. As to the ratio of volumes in which the two acids are to be mixed, it may vary to a considerable extent, and the specific gravity may vary also; but the above is preferable. The ratio, however, yielding the best results is that of one volume of nitric acid to two or three of sulphuric acids.

Second. Into a mixture of the acids of the last-named description clean cotton is plunged in such a manner as to be entirely covered and impregnated by the said acid mixture. Care must be taken that the temperature of the said acid mixture be not above 50° or 60° Fahrenheit.

Third. The cotton, after having remained for an hour or two (more or less) in contact with the acid mixture, must be subject to pressure in order to remove as much as possible from the cotton the acid particles. That being done, the pressed cotton is to be washed as long until the acid is entirely removed. That operation being finished, the prepared cotton is to be dried in moderately-heated rooms. Before using the dry cotton it is important to card it.

Fourth. Cotton-wool acquires also a high inflammability and explosive power by exposing that material at the common temperature to the action of pure nitric acid of the greatest specific gravity that acids can be prepared. This way of rendering the cotton inflammable appears to be less easy and economical than the method above described; but the use of this cotton-wool so prepared I claim as part of my discovery.

Fifth. For many purposes it is good to impregnate the explosive cotton by some nitrate of potash. This prepration imparts to the cotton the property of disengaging a more intense light than the pure prepared cotton does, and the use of nitrate potassa or other known chemical substitutes I claim in combination with the acid treatment.

In testimony whereof I, the said CHRISTIAN FREDERICK SCHÖNBEIN, herewith subscribe my name in the presence of the consul of the United States of America.

CHRISTIAN FREDERICK SCHÖNBEIN.

Witnesses:

JAMES M. CURLEY,
JOSEPH MARQUETTE,
Clerks in the Consulate of the U. S., London.

Patent for the "Improvement in Preparation of Cotton-Wool and Other Substances as Substitutes For Gunpowder"

achievements was his discovery and naming of ozone, but he will best be remembered for his work on nitrocellulose and his suggestion that this material be substituted for gunpowder. There is a widely told story that guncotton was discovered in his kitchen. He often did his experimenting at home while his wife was out of the kitchen. One day while he was distilling nitric and sulfuric acids on the stove, his flask broke. He grabbed the nearest thing, his wife's cotton apron, and wiped up the mess. He carefully washed the apron in water and hung it up to dry. Shortly afterward, he heard an explosion and saw the apron instantly vanish, leaving no smoke or ash. Whether or not this story is true, the fact remains that Schönbein did discover guncotton and recognized its military importance.

Due to political conditions, he was unable to sell his secret to Germany, but Austria was more receptive. In 1846, he sold the formula to England where it was patented in the name of John Taylor and issued as British Patent 11,407 on October 8, 1846. He entered into an agreement with John Hall & Sons that gave them the sole right to manufacture guncotton at their powder works at Faversham. By December 1846, not only Austria and England, but also France, Germany, and Russia were building guncotton plants. However, the manufacture of guncotton was severely curtailed when in July 1847 the plant at Faversham exploded with the loss of 21 lives. In quick succession, similar explosions occurred in Russia, Germany, Austria, and France. Guncotton manufacture was abandoned for several years until Dewar and Abel found methods to tame it.

Christian Friedrich Schönbein died unexpectedly at Sauersberg, Baden, on August 29, 1868, and was buried at Basel, Switzerland.

Gyroscope

Ship and aircraft stabilization was made possible by the invention of the gyroscope. One of the early inventors in the field was Elmer A. Sperry, who received the first U.S. patent, No. 1,279,471, for a gyrocompass on September 17, 1918.

The gyroscope is an apparatus consisting of a rapidly spinning heavy wheel that maintains its original position when its axis of rotation is moved. When used for stabilizing ships and aircraft, it is known as a gyrostabilizer. When used as a compass, it is known as a gyrocompass.

In 1740, Leonhard Euler, a Swiss mathematician, worked out the equations of motion for the gyroscope. According to written records, the first gyroscope was built by G. C. Bohnenberger of Germany in 1810. In 1836, Edward Sang of Scotland proposed the idea of using the gyroscope to demonstrate the rotation of the earth but lacked the resources to construct an accurate rotor. In 1852, French physicist Jean Bernard Leon Foucault used a hand-driven gyroscope to detect the rotation of the earth. He also gave it the name "gyroscope." The success of Foucault was due to a large extent to the mastery with which Gustave Froment of the Ecole Polytechnique in Paris constructed the gyroscope that Foucault had designed. In 1857, the first commercially manufactured gyroscopes in America were made by the Holbrook Apparatus Company, Hartford, Connecticut. They were made of iron, sold for two dollars, and were used as toys. In 1868, Matthew Watt-Boulton of England first proposed the use of gyroscopes to stabilize ships. In 1896, Robert Whitehead, a Scottish engineer, first used the gyroscope in military equipment to control a torpedo. In 1903, Herman Anschutz-Kaempfer of Germany patented the first gyrocompass and, in the same year, Otto Schlick of Germany devised the first gyrostabilizer for a ship. In 1918, Elmer Sperry, an American, not only received the first U.S. patent for a gyrocompass but also designed and manufactured such superior gyroscopes that the American Navy, as well as the British Navy, adopted them for their use.

Elmer Ambrose Sperry was born in Cortland, New York, on October 12, 1860. His mother died soon after his birth, so he was raised by his paternal grandparents. His early education was completed in Cortland, and, as a teenager, he worked in a book bindery after school hours. He attended the State Normal Training School at Cortland for three years and spent one year in 1878 at Cornell University. In 1876, with the help of the Young Men's Christian

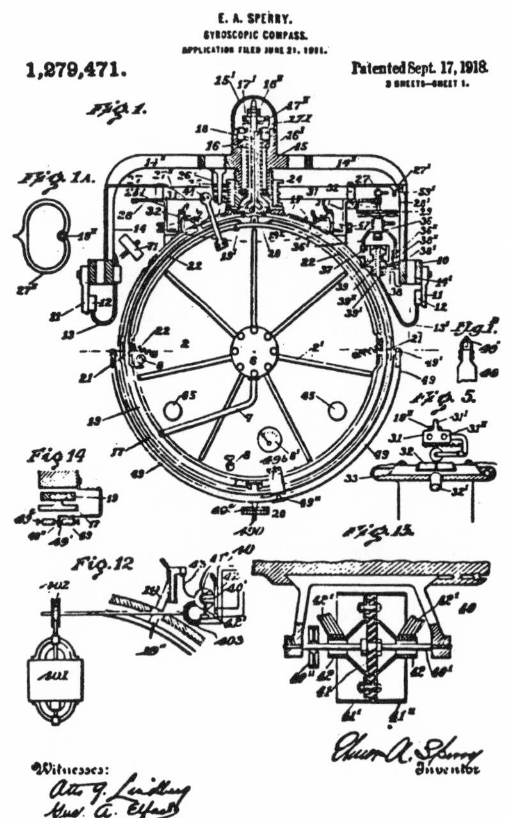

Elmer A. Sperry's gyroscope patent, issued September 17, 1918.

Association of Cortland, he visited the Centennial Exhibition in Philadelphia where the mechanical exhibits made an impression on his life.

In 1879, with the financial help of a Cortland manufacturer, he constructed an improved dynamo and arc lamp. In the same year, he went to Syracuse, New York, and built a large dynamo capable of operating a series of arc lamps. In 1880, he founded the Sperry Electric Company in Chicago, Illinois, to manufacture dynamos and arc lamps. In a short time, this company became extremely successful.

In 1888, he organized the Sperry Electric Mining Machine Company in Chicago to manufacture mining equipment for use in soft coal mines. In 1890, he founded the Sperry Electric Railway Company and built a plant in Cleveland, Ohio, to manufacture electric street-railway cars. In 1894, he sold the plant, along with his patents on street railway machinery, to the General Electric Company. In 1894, he became interested in electric vehicles, particularly the storage battery. He improved and patented an improved storage battery that was capable of operating a vehicle over 100 miles. He founded the National Battery Company and built a factory at Buffalo, New York, to produce his storage battery.

In 1900, he established a research laboratory in Washington, D.C., to do research on electrochemistry. His associate was C. P. Townsend, an electrochemist. They developed the Townsend Process for manufacturing pure caustic soda from salt. They also developed the chlorine detinning process for recovering tin from old cans and scrap. This process was eventually taken over by the Goldsmith Detinning Company. Sperry organized the Chicago Fuse Wire Company to produce electric fuse wire using a machine that he had devised.

Sperry began working on a gyrocompass in 1896. By 1910, he had perfected it enough to establish the Sperry Gyroscope Company in Brooklyn, New York. In 1911, his first gyrocompass was tried out on the battleship Delaware at the Brooklyn Navy Yard. Together with his sons, Lawrence and Elmer Jr., he built gyrostabilizers for ships in 1913 and gyrostabilizers for aircraft in 1914. Son Lawrence entered a contest in Paris in 1914 for the most stable airplane and won the prize of 50,000 francs over 53 other contestants. Sperry's other inventions that use the principle of the gyroscope include the automatic pilot for airplanes, which accompanied Wiley Post on his flight around the world; the gyro pilot which automatically steers a vessel upon a prescribed course; the gyro-roll and pitch recorder, which permits a comparison to be made between the motion of a vessel with and without the stabilizer; and the gyro track recorder, which enables the irregularities in a railroad track to be located and measured. In 1929, he disposed of his interest in the Sperry Gyroscope Company. His son Elmer Jr. continued the work and invented the directional gyro and artificial horizon, which tells a pilot the direction as well as the tilting and turning of the plane and also tells the true position of the natural horizon and whether the plane is climbing or diving. With this equipment, Jimmy Doolittle made the first covered-cockpit flight and landing.

In 1918, Sperry produced a high-intensity arc searchlight that was 500 percent brighter than any previous light. This light made possible the making of indoor motion pictures.

The inventor was awarded over 400 patents in the United States and Europe. He founded eight companies to manufacture his inventions. He received the John Scott Legacy Medal and Premium in 1914 from the Franklin Institute, the Collier Trophy in 1915, the John Fritz Medal in 1927, the Holley Medal in 1927, the American Iron and Steel Medal in 1929, and the Elliott Cresson Medal from the Franklin Institute in 1929. He received two decorations from the Emperor of Japan and two decorations from the Czar of Russia. Stevens Institute of Technology, Lehigh University, and Northwestern University gave him honorary degrees.

Sperry died June 16, 1930. He never forgot the Y.M.C.A. of Cortland, New York, for sending him to the Philadelphia Centennial Exhibition, and in his will he left $1 million to the national organization.

Halftone Printing Plate

The reproduction of photographs in newspapers and magazines was made possible by Frederic Eugene Ives who received the first U.S. patent, No. 237,664, on February 8, 1881, for the halftone printing screen and process.

The printing of pictures was first accomplished by hand carved woodcuts. During the Civil War, thousands of detailed engravings were made by this method for periodicals and books. In 1852, Henry Fox Talbot of England first suggested a photographic means for reproducing pictures by interposing a line screen to divide the photograph into lines. For this, he received British Patent No. 565 on October 29, 1852. In 1871, Carl Gustaf Wilhelm Carleman, a Swedish engraver, used the first successful printer's ink single-line halftones to illustrate his book, *Photography By Typographic Printing Press*. On March 4, 1880, Stephen Horgan, an American, published the first halftone picture in an American newspaper, the *New York Daily Graphic*, by using a screen on which parallel lines had been carefully drawn. The source was from Henry J. Newton's photograph of New York's Shantytown. Frederic Eugene Ives, an American, conceived the basic idea of the modern halftone screen in 1881. He cemented together, face to face, two line screens with the lines placed perpendicular to each other. By 1891, screens of this type were being commercially produced in Philadelphia by Louis and Max Levy.

Frederic Eugene Ives was born February 17, 1856 in Litchfield, Connecticut. His father, Hubert Leverit Ives, was a farmer but gave up farming in 1866 when he contracted tuberculosis. He moved his family to Norfolk, Connecticut, where he became a village storekeeper. It was in Norfolk that young Frederick discovered a book on natural philosophy in an attic storeroom, and purchased a Youth's Companion Premium Microscope for a dollar. In 1868, his father died, and he went to work as a clerk in another store where he found a small hand printing press. These three items—the book, the microscope, and the printing press—had a great influence on his life.

When Ives was 14 years old, he became an apprentice in the printing shop of the *Litchfield Enquirer* in Litchfield, Connecticut. There, he taught himself the art of wood engraving and photography, making his first photographic negative by the collodion and silver-bath process from instructions found in a Baltimore Photo-Catalog. He soon became an expert photographer and in 1873, after completing his apprenticeship, went to work in a printing shop in

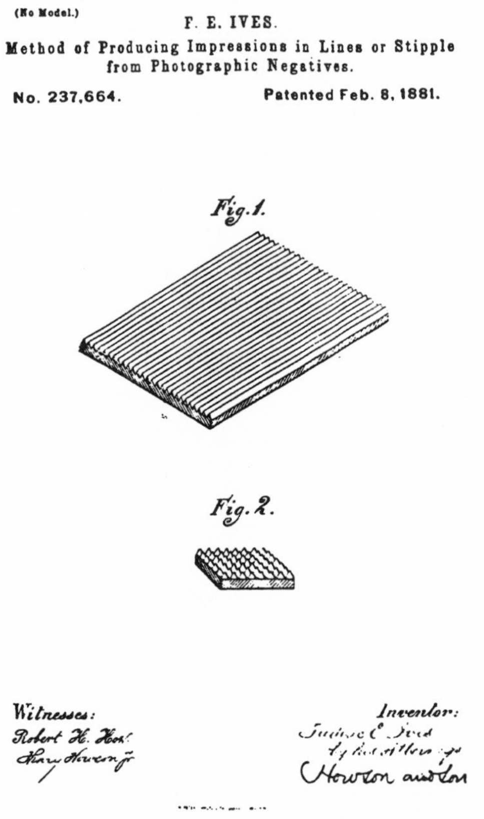

(No Model.)

F. E. IVES.

Method of Producing Impressions in Lines or Stipple from Photographic Negatives.

No. 237,664. Patented Feb. 8, 1881.

Fig. 1.

Fig. 2.

Witnesses: Inventor:

F. E. Ives' "Method of Producing Impressions in Lines or Stipple from Phtographic Negatives"

Ithaca, New York. Soon after, he opened his own photographic studio in Ithaca and in 1875 became a photographer in the photographic laboratory of Cornell University. A year later, he developed a photo-mechanical process of making printing plates from pen drawings using "swelled gelatin," which he used for two years in illustrating the college paper, *Cocagne*. In 1878, he made a gelatin relief from a photographic negative and then by mechanical means converted this relief into dots of varying sizes from which a printing plate could be made. Thus, the halftone method of producing printing plates was born.

In 1879, Ives moved to Philadelphia and became associated with the firm of Crosscup and West. He established a photoengraving department at this firm and continued to improve his halftone process. The first commercially used halftone by the Ives process appeared in the June 1881 issue of the *Philadelphia Photographer*. In 1885, Ives sealed together two lined screens face to face with the lines perpendicular to each other and created the modern halftone screen method for producing halftone printing plates.

In 1885, Ives began to experiment with color photography and three-color printing. He exhibited a color reproduction made by three halftone plates at the Novelties Exhibition in Philadelphia. This exhibit was so well received that he turned his attention completely to color photography. In 1890, he developed the photochromoscope, called the Kromskop. This device enabled him to make three negatives of the same object through a single lens and then, after converting to positives, to simultaneously project these positives through color filters to yield the original object in full color. In 1903, he invented and patented his parallax stereogram, the forerunner to 3-dimensional photography. In 1912, he developed the Tripak color camera, in 1916 the Hi-Cro color process, and in 1928 the two-color process, Polychrome.

Ives received a total of 70 patents but many of his inventions were not patented. He received many honors and awards. The Franklin Institute alone awarded him ten medals, including the Elliott Cresson Medal. He also received medals from the Photographic Society of Philadelphia, the Society of Arts of London, the Photographische Gesellschaft of Wien, the Royal Scottish Society of Arts of Edinburgh, the Royal Photographic Society of London, and the American Academy of Arts and Sciences.

Ives died May 27, 1937, at his home in Philadelphia and was cremated at the Chelton Hills Crematory. His son, Herbert Eugene Ives, also distinguished himself in the art of telephotography by transmitting pictures by telephone wire in 1924.

Helicopter

The first U.S. patent for a successful single-rotor helicopter was granted to Igor Ivan Sikorsky in 1932. The number was 1,848,389, issued March 8, 1932.

The first recorded design for a helicopter was made by the Italian artist and scientist Leonardo da Vinci in 1483. These drawings introduced the direct-lift principle of flight. This idea lay dormant for many years until the French mathematician, Paucton, in 1768 designed a direct-lift machine, which he called a Pterophone. The machine had a horizontal rotor for lift and a vertical rotor for thrust, with both rotors to be driven manually by the pilot. In 1784, M. Launoy and M. Bienvenu, both of France, demonstrated at the Paris World's Fair a toy helicopter that rose to a height of 70 feet. It had two four-bladed rotors made of turkey feathers and spring actuated that rotated in opposite directions. It is claimed to be the first heavier-than-air powered craft to fly.

In 1842, Horatio Phillips of England was the first to build a model jet-propelled helicopter that used steam jets at the rotor tips. In 1843, Sir George Cayley of England, later known as the "father of British aeronautics," was the first to design a full-scale helicopter. It was steam powered and known as the Aerial Carriage. In 1859, the first British patent for a helicopter, No. 2,330, was awarded to Henry Bright of England.

Viscount de G. L. M. Ponton d'Amecourt of France was the first to use aluminum in the construction of a helicopter. This is stated in British Patent No. 1,929 of 1861. In 1877, Enrico Forlanini, an Italian engineer, built a contra-rotating, superheated steam-powered model helicopter that rose to a height of 43 feet and remained air-borne for 20 seconds. In 1904, Charles Renard of France built a twin-rotor helicopter using a gasoline engine as the power source. It flew by itself, but without a pilot. In 1905, Emile Berliner of the United States, and inventor of photographic records, built an 80-horsepower co-axial rotor helicopter that succeeded in lifting its own weight. His son Henry continued in this work and designed a machine that flew with some success in 1924.

In 1907, Louis Charles Breguet of France built a square, full-sized helicopter that had four rotors. It was called a Helicoplane but was not successful, since it had no devices to control it in flight. In the same year, Paul Cornu, a French mechanic, built a full-sized helicopter with rotor blades that looked

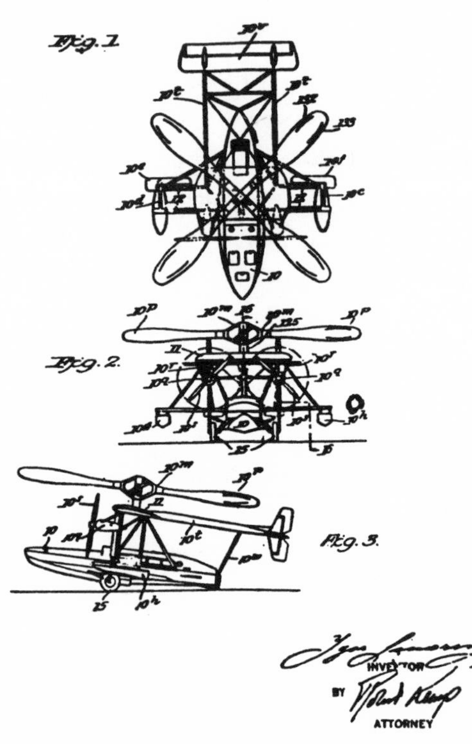

Sikorsky's "Aircraft of the Direct Lift Amphibian Type and Means of Constructing and Operating the Same"

like giant fly-swatters. This helicopter was the first to leave the ground carrying its own pilot. In 1909, Igor Sikorsky of Russia built a 25-horsepower helicopter that had contra-rotating rotors on concentric shafts, one shaft revolving inside the other. Also in 1909, the first U.S. patent, No. 1,021,521, for a helicopter was granted to P. L. T. Heroult of France.

In 1923, Juan de la Cierva of Spain made the first autogiro, a machine that combined the features of a conventional airplane with those of a helicopter. His was the first to have flexible rotors. In 1925, A. G. Von Baumhauer of Holland built the first single-rotor helicopter with a tail rotor driven by a separate motor.

In 1923, the United States Army Air Force asked George de Bothezat, an American aviation scientist, to build a military copter. When finished, his machine, called the *Flying Octopus,* had four large rotors mounted at the ends of a fuselage built in the shape of a cross. It had stable flight, carried over 4,000 pounds and three passengers, and cost the army $200,000. After watching over 100 flights, however, the military men rejected the machine saying it was too hard to fly. Heinrich Focke of Germany designed the first practical helicopter with two, counter-rotating rotors arranged side by side in 1937. It was called the Focke-Achgelis FW-61 and set many records. However, the world did not learn of this helicopter until after World War II.

The military again became interested in helicopters in 1942 and asked Igor Sikorsky, now a successful designer and manufacturer of airplanes in Connecticut, to build one that would meet military standards. In May 1942, he delivered to the U.S. Army at Wright Field, Dayton, Ohio, its first helicopter, the Sikorsky VS-300. This was followed by the 180-horsepower Sikorsky XR-4. These machines were the first successful single-rotor helicopters and firmly established Sikorsky's name among the great in aircraft development.

Igor Ivan Sikorsky was born May 25, 1889, in Kiev, Russia. His father was professor of psychology at St. Vladimir University in Kiev, and his mother was a medical school graduate. From them, he acquired an early interest in science and before he was even a teenager, he had built batteries, an electric motor, chemical bombs, and a steam-powered motorcycle. After viewing the drawings of Leonardo da Vinci, he developed an interest in helicopters, an interest that would remain with him for the rest of his life.

In 1906, he graduated from the Imperial Naval College in St. Petersburg and, for the next two years, attended the Mechanical College of the Polytechnic Institute in Kiev. In 1909, he borrowed some money from his older sister Olga to go to Paris to study aeronautics and to buy a gasoline engine for a helicopter that he had designed a year earlier. Four months later, he returned to Kiev with a three-cylinder, 25-horsepower engine built by a Mr. Anzani, a designer of racing motorcycles.

In May 1909, he started to work on his helicopter. He built a large wooden box with the engine on one side and the pilot's seat on the other. He rigged pulleys and shafts to transmit the power of the engine to a couple of two-bladed rotors 16 feet in diameter, one on top of the other, which rotated in opposite directions. In a test run, he found that his machine had a lifting capacity of 350 pounds, but it weighed 450 pounds. He built a second helicopter having a butterfly design, but it was no more successful in flight than the first.

After 18 months, Sikorsky turned his attention to the conventional fixed-wing airplane. He completed his first airplane, the S-1, in 1911, and it flew about four feet above the ground for 12 seconds. He built another, with improvements, but after a few takeoffs and landings, it crashed and was wrecked beyond repair. With the help of his family, he made another trip to Paris and returned with a 40-horsepower Anzani engine. In three months, he built the S-3, which performed without any trouble until the 13th flight. On this flight, the engine lost power forcing Sikorsky to land on a frozen pond that couldn't bear the weight of the plane and pilot. With some personal loans from friends and a mortgage on the

Sikorsky's house, he built the S-4 and the S-5 in 1911. With the S-5, he set a world's record when the machine carried three men on a 30 mile cross-country flight at a speed of 70 miles an hour. Success had finally come.

He was invited to participate in the maneuvers of the Russian Army near Kiev with his S-5. After this successful mission, he was made chief engineer and designer in 1912 of the aviation department of the Russo-Baltic Railroad Car Works in Petrograd. In 1913, he constructed the Grand, the first four-motored plane in history. It weighed 9,000 pounds, had a span of 92 feet, four 100-horsepower engines, a decorated 16-passenger cabin, two pilots' seats with double control, and a toilet. This plane made aviation history in 1913 and 1914 when 75 of these four-engine planes were converted into bombers and used by the Russian Army to carry out more than 400 air-raids on German positions and supply bases during World War I.

The Russian Revolution of 1917 caused Sikorsky to leave his native Russia. After a short time in Paris and London, he arrived in New York in March 1919. Unable to get a job in the aircraft industry, he made a living as a mathematics teacher at a school for Russian immigrants on New York's East Side. At night, he drew sketches and worked out designs for a new airplane. With the help of friends buying shares at $10 each, a share company, the Sikorsky Aero Engineering Corporation, was formed in 1923. One of these friends was Serge Rachmaninoff, the famous pianist and composer, who subscribed $5,000. A factory was built on a farm near Roosevelt Field, Long Island, and work was begun on the construction of the S-29, an all-metal, twin-engined airliner. This 14-passenger plane with a speed of 115 miles an hour was a complete success and enabled Sikorsky to re-establish himself as a leading aircraft designer.

He organized the Sikorsky Manufacturing Corporation in 1925 and reorganized it as the Sikorsky Aviation Corporation in 1928. It became a subsidiary of United Aircraft Corporation in 1929. A new factory was built in Bridgeport, Connecticut, and it was from this factory that the American Clipper was built in 1931, the first of a series of four-engined Clipper ships which established regular transoceanic air service.

In 1939, Sikorsky renewed his interest in the helicopter. He built his famous VS-300 that was followed by his equally famous XR-4 in 1941. The U.S. Army purchased its first helicopter from Sikorsky in 1942. During World War II, the only helicopters flown by the Air Force were Sikorsky helicopters.

Many honors came to Sikorsky. Czar Nicholas II of Russia awarded him a gold watch in 1913. He became a naturalized United States citizen in 1928. Yale and Lehigh Universities conferred honorary degrees on him. He was awarded the Potts Medal of the Franklin Institute in 1933, the Hawkes Memorial Trophy in 1947, the Daniel Guggenheim Medal in 1951, and the National Defense Transportation Award in 1952. He was the author of two religious books, *The Message of the Lord's Prayer* in 1942 and *The Invisible Encounter* in 1947.

He died in Easton, Connecticut, on October 26, 1972.

Hydroplane

The first successful airplane capable of landing and taking off from the water was the invention of Glenn Hammond Curtiss. For this accomplishment, he was awarded the first United States patent for a hydroplane. The application was filed on August 22, 1911, and matured into United States Patent No. 1,420,609 on June 20, 1922.

For years individuals had dreamed of flying through the air like a bird. Many devices were built to fulfill this dream. These include balloons, gliders, dirigibles, helicopters, and airplanes. History records that in 1783, Joseph and Etienne Montgolfier of France constructed a model balloon 30 feet in diameter, filled it with hot air and smoke, and watched it ascend and float about a mile away. In 1852, Henri Gifford of France built a steam-powered, propeller-driven dirigible and achieved man's first controlled flight. In 1853, George Cayley of England built and had his coachman fly the first glider. In 1903, Orville and Wilbur Wright of Dayton, Ohio, built and successfully flew the first motor-driven airplane in pilot-controlled flight. In 1907, Paul Cornu of France built a full-sized helicopter and was the first to leave the ground carrying its own pilot.

These devices had one thing in common. They required the ground as the landing and taking off point. In 1911, Glenn Curtiss further advanced the field of aviation by being the first to take off and land an airplane on water.

Glenn Hammond Curtiss was born on May 21, 1878, in Hammondsport, New York. His father, Frank R. Curtiss, a harness shop owner, died when Glenn was only six years of age. His mother moved the family to Rochester, and young Curtiss spent his early years between his mother and his paternal grandmother in Hammondsport. His early schooling revealed that he had an aptitude for machinery and mathematics. After grade school, the boy found a job in Rochester as a messenger boy in a telegraph office. It was there that he developed a love for bicycles. He

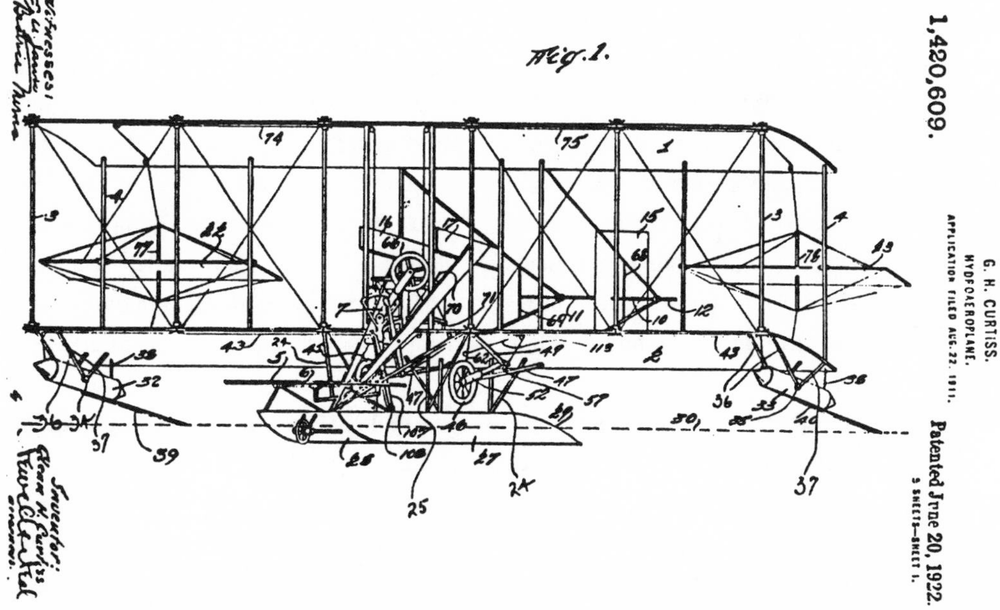

Glenn Hammond Curtiss's Hydroaeroplane, filed in August of 1911, eight years after Orville and Wilbur Wright's successful motor-driven flight.

returned to Hammondsport in 1898 and opened a bicycle shop of his own with the help of a local businessman. There, not only did he repair, build, and sell bicycles under his trademark, Hercules, but he became an excellent bicycle racer, winning every race in nearby competitions.

In 1900, motorcycles appeared on the American scene, and these speedier vehicles captured the interest of Curtiss. He made improvements in the engines and frames and in 1902 established the G. H. Curtiss Manufacturing Company to make and sell motorcycles. As with bicycles, he not only sold motorcycles but raced them as well. In 1903, he won the national championship race at the Empire State Track in New York City.

Curtiss was attracted to aviation when Thomas Scott Baldwin, a dirigible pioneer, ordered an engine for one of his balloons. A short time later, Baldwin moved his balloon plant to Hammondsport where he and Curtiss built the first army dirigible. Curtiss was not a dirigible enthusiast because of the craft's slow speed, but his company did enjoy considerable success building engines for balloonists.

Curtiss became interested in airplanes after meeting Alexander Graham Bell, inventor of the telephone. Bell, a follower and admirer of Samuel Pierpont Langley, began to experiment with heavier-than-air flight after Langley abandoned his efforts. Bell enlisted Curtiss in 1905 to build an engine for his experiments. The first engine built by Curtiss was unsatisfactory. The second engine was successful and was delivered to Bell in person by Curtiss in Baddeck, Nova Scotia, where the experiments were being conducted. From this point on, Curtiss spent the rest of his life designing and building airplanes.

In 1907, Bell, along with other supporters, founded the Aerial Experiment Association, which was headquartered at Hammondsport. Curtiss was named as director of experiments. On July 4, 1908, the first publicized flight in the United States took place at Hammondsport in a plane designed by Curtiss, called the *June Bug*. The flight won for the designer the *Scientific American* trophy. In August 1909, he attended the First International Aviation Meet in Rheims, France. His competition was Louis Bleriot, a French aviator who was the first to fly across the English Channel. Curtiss won the race by six seconds and was awarded the Gordon Bennett Cup. After this event, he was honored as a hero in both Europe and the United States.

Curtiss believed that an airplane could be modified to land and take off from water. He placed pontoons on one of his planes and on January 26, 1911, he succeeded in taking off and landing on the water of San Diego Bay. He had invented the hydroplane. The United States, as well as many foreign countries, purchased many of these planes. In 1912, he invented the flying boat, and in 1914 built the first flying craft designed for transatlantic flight for Lewis Rodman Wanamaker, the department store magnate. This plane never made the trip, but a Curtiss-built flying-boat was the first to cross the Atlantic on May 27, 1919. During World War I, the Curtiss Aeroplane & Motor Company was founded which made more than 5,000 Jennies, the Curtiss standard airplane. After the war, his presence at his airplane companies diminished, and he became interested in real estate near Miami, Florida. However, his passion for research continued, and in 1929 he developed a streamlined trailer for luxurious traveling.

Curtiss died on July 23, 1930 in Buffalo, New York, from complications of appendicitis. He was buried in the Pleasant Valley Cemetery, Hammondsport, New York. His funeral was attended by aviation experts and government officials from all over the world.

In 1913, the Smithsonian Institution awarded him the Langley Metal for his development of the hydroplane. In 1933, the United States government awarded him posthumously the Distinguished Flying Cross for his contributions to American aviation.

Incandescent Lamp

One of the truly remarkable inventions of our time is the incandescent lamp. This was the result of the work of Thomas Alva Edison, who invented the first successful lamp and received U.S. Patent No. 223,898 on January 27, 1880.

Some 19 electric lamps were invented before 1880. These inventors include James Bowman Lindsay of Scotland, who developed the first practical electric lamp on July 25, 1835; William Robert Grove, inventor of the Grove battery, in 1840; J. W. Starr, who received a U.S. patent in 1845; August King, who was granted British Patent No. 10,919 in 1845; Goebel of Germany, who in 1850 made several battery-operated lamps consisting of evacuated glass tubes with thin carbon filaments; S. Gardiner and L. Blossom, who received U.S. Patent No. 20,706 on June 29, 1858; Moses G. Farmer, who lit his home in Salem, Massachusetts, in 1859 and who was granted U. S. Patent No. 213,643 on March 25, 1879; Joseph Wilson Swan, who made an electric lamp with a carbon filament in 1860, and who later joined forces with Edison in forming the Edison and Swan United Electric Light Company Limited in 1883; Sawyer and Man, who received U.S. Patent No. 205,144 on June 18, 1878; and St. George Lane-Fox, who received British Patent No. 3,988 on May 14, 1879. However, it was Thomas Alva Edison who perfected a practical incandescent lamp that became a worldwide industrial and household item.

Thomas Alva Edison was born in Milan, Ohio, on February 11, 1847. The family moved to Port Huron, Michigan, where Edison received only three months of formal schooling. His mother, a former teacher, gave him an adequate elementary education at home. An attack of scarlet fever impaired his hearing, which became progressively worse with time.

At the age of 10, his favorite study was chemistry and at the age of 11, he was experimenting in his father's cellar with chemicals obtained from the village drug store. At the age of 12, he ran a news and vegetable stand in Port Huron and got a concession on the Grand Trunk Railway between Port Huron and Detroit, where he was allowed to have a small laboratory and printing press in the baggage car. In addition to his chemical experiments, he published a weekly journal, called *The Weekly Herald.* A baggage car fire, started by one of his chemicals, ended his train experiments and journal when Edison, his chemicals, and printing press were thrown off the train by the conductor. At the age of 15, he was taught Morse code as a reward for saving the life of a station agent's baby, and after three months training, he got a job as a telegraph operator.

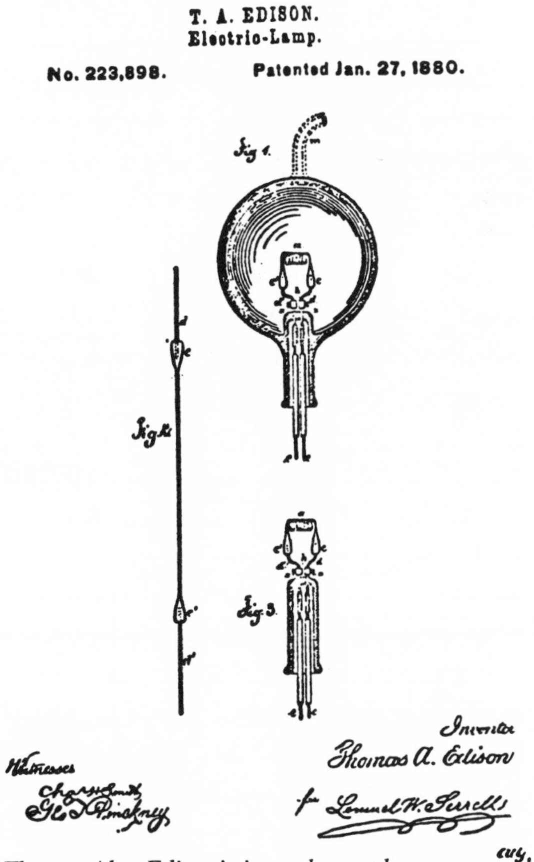

T. A. EDISON.
Electric-Lamp.
No. 223,898. Patented Jan. 27, 1880.

Thomas Alva Edison's incandescent lamp

From 1863 to 1868, Edison worked as a wandering telegrapher, his last telegraphic job being with Western Union in Boston. It was there that he started work on his first patented invention, an electric voting machine. After leaving Western Union, he became a free-lance inventor in Boston, where he invented the duplex telegraph and worked to improve stock tickers.

Edison then went to New York and eventually became general manager of the Gold Indicator Company. After a series of mergers, this company was absorbed by Western Union. Not wishing to work for Western Union again, Edison quit to become a free-lance inventor once again. General Marshall Leffersts, the president of Western Union, offered Edison $40,000 for the rights to his inventions. Edison accepted and opened a shop in Newark, New Jersey, with a staff of 50 men to manufacture stock tickers. This left him time to work on his inventions. It was there that he invented waxed wrapping paper, the electric pen, and the mimeograph.

In 1876, he bought an estate at Menlo Park, New Jersey, and moved his laboratory from Newark. There, he devoted his entire time to inventions. Edison had a large staff that, from the beginning, included names that became famous in electrical engineering. At one time or another on Edison's payroll were Schuckert, the founder of the Siemans-Schuckert Works in Germany; John Kruesi, who became the chief engineer of General Electric; Arthur Kennelly, who discovered the Kennelly-Heaviside layer; Edward Acheson, who later invented Carborundum; and John Fleming, who was to invent the radio tube.

It was at the Menlo Park estate that Edison invented the incandescent light. He began experimenting with various metallic filaments but became convinced that the only material that would serve his purpose was a thin thread of carbon. He had noticed that strips of carbonized paper glowed for a few moments with a bright, incandescent light when an electric current was passed through them. He searched for a more suitable material and after 13 months of experimenting and the expenditure of thousands of dollars, he discovered that carbonized, ordinary sewing-thread was the best solution. So in 1879, he carbonized a cotton thread, placed it in a globe from which the air had been pumped, sealed it, and passed an electric current through two pieces of platinum wire that had been cemented to the cotton thread. The globe glowed for 40 hours. It was the first practical incandescent electric lamp.

Edison said later, "We sat and looked and the lamp continued to burn. The longer it burned the more fascinated we were. None of us could go to bed, and there was no sleep for 40 hours. We sat and just watched it with anxiety and glowing elation."

Public announcement of the invention was made in the *New York Herald* on December 21, 1879. Edison demonstrated his electric lamps to the world by illuminating his own laboratory at Menlo Park with 500 bulbs in 1880. Extra trains were run from New York, and engineers crossed the Atlantic from Europe to view the demonstration.

The first commercial water installation was on the steamship Columbia of the Oregon Railway and Navigation Company. The first commercial land installation was in 1881 in the lithographing shop of Hinds, Ketchum, and Company, 229 Pearl Street, New York. The first public-service station was put into operation at Appleton, Wisconsin, in 1882, where Edison installed one of the world's first hydroelectric plants. It was powered by the Fox River and lit between 200 to 300 lamps in the neighborhood. Also in 1882, central stations were established at Pearl Street in New York, Holborn Viaduct in London, Sunbury in Pennsylvania, and in Milan, Italy.

Edison filed his first patent application at age 21 and received the patent to his incandescent lamp at age 32. During his 60 years of experimentation, he received 1,097 U.S. patents and 1,239 foreign patents. He became the first member of the Inventors Hall of Fame in Arlington, Virginia. President Nixon signed a resolution establishing Edison's birthday on February 11 as National Inventors Day. Edison died on October 18, 1931.

Instant Photography

The first United States patent for a one-step photographic process was granted to Edwin Herbert Land on February 27, 1951. The number was 2,543,181.

The first practical photographic method for producing a picture was introduced by Louis J. M. Daguerre, a French inventor and painter, in 1839. He made a polished, silvered copper plate sensitive to light by subjecting it to iodine vapors. He then exposed it from 3 to 30 minutes in a camera. He developed the image with mercury vapor and fixed it with sodium thiosulfate.

Also in 1839, British scientist William H. F. Talbot invented the negative-positive process for making photographs. The inventor made a sheet of paper sensitive to light with sodium chloride and silver nitrate solutions, then exposed it in a camera where the silver turned dark when exposed to light. Using a salt solution, he then produced a negative from which many paper positives could be produced.

In 1851, Frederick Scott Archer, an English sculptor, introduced the wet-collodion process. He coated a glass plate with collodion that had been combined with potassium iodide. He then dipped the plate in a solution of silver nitrate. While still wet, the plate was exposed in a camera, immediately developed with gallic acid, and fixed with hypo.

In 1871, Richard L. Maddox, an English physician, discovered the dry-plate process. He replaced the collodion with gelatin and the silver iodide with silver bromide. He coated a glass plate with this gelatin-silver bromide emulsion and allowed it to dry. This plate could then be camera exposed at a later date and developed whenever the photographer chose.

These processes involved much time and many steps. Most of them included exposing, removing the film from the camera, developing, and fixing the image in a darkroom to form a negative, and then making a positive from the negative. In 1947, Edwin H. Land reduced the time to less than a minute and the process to one step when he introduced his instant photographic process.

Edwin Herbert Land was born in Bridgeport, Connecticut, on May 7, 1909. His father, Harry M. Land, was in the scrap-metal and salvage business. The family moved to Norwich, Connecticut, where young Land attended Norwich Academy and later enrolled in Harvard University. While at Harvard, he became interested in light polarization and spent

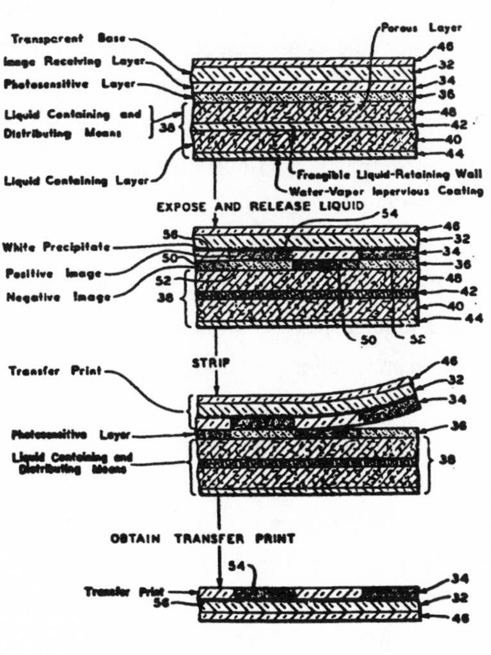

Instant Photography began with this 1948 patent, awarded to E. H. Land

many hours experimenting in this field in a laboratory provided by the university. In 1932, he succeeded in making a light-polarizing film by dipping a sheet of polyvinylalcohol in an iodine solution. He named this substance Polaroid, the first synthetic light polarizer.

Land and George Wheelwright III, a physics instructor, left Harvard in 1932 and established a consultant service in Boston called Land-Wheelwright Laboratories. In 1935, they began to manufacture polarizing filters for Eastman Kodak Company to use in cameras. In 1936, the American Optical Company began using Polaroid in sunglasses. Increased business resulted in establishing the Polaroid Corporation in 1937 in Cambridge, Massachusetts, with the financial backing of such men as W. Averell Harriman and Lewis Strauss. The Polaroid Corporation purchased the Land-Wheelwright Laboratories.

During World War II, Land worked on many government projects including the National Defense Research Council. During this time, he invented many improvements on weapons particularly in the optical field. The Polaroid Corporation contributed to the war effort by manufacturing goggles, gun sights, periscopes, range finders, aerial cameras, and the Norden bombsight. Land, together with Joseph Mahler, developed a film for three-dimensional pictures called Vectographs. These were used during the war and later in X-ray work and 3-D movies.

While on vacation with his family in 1943, Land conceived the idea for a camera and film that would produce an immediate photograph. After he took a picture of his three-year-old daughter Jennifer, she asked why she could not see the picture right away. Within an hour, Land answered his daughter's question, at least in his own mind, by clearly seeing the camera, the film, and the physical chemistry. However, it was another three years before these ideas were made practical.

On February 21, 1947, Land introduced his one-step camera and film before the American Optical Society in New York. In this process, the camera became the darkroom. A full-sized silver halide negative was exposed and brought into contact with a positive print sheet by rollers. These rollers ruptured a pod of developing reagents attached to the film that spread evenly between the negative and the positive and developed these simultaneously. After a minute, the positive picture was separated from the negative. This system was an immediate success though only sepia prints were developed. On November 26, 1948, the first Polaroid Land Camera went on sale at the Jordan Marsh department store in Boston. The first year, sales soared to about $5 million. In the next 11 years, Polaroid obtained about 300 patents in the photographic field, 120 of which were issued to Land. One of these was for a black and white film to replace the sepia print.

In 1947, Land started thinking about instant color film and asked Howard G. Rogers, a Polaroid chemist, to work on the problem. Rogers eventually came up with dye developers, which were a combination of preformed dye and developer. By 1957, Rogers had made a prototype of the first one-step color photograph. This film was introduced to the market in 1963 under the name Polacolor, which, like the black and white film, was a peel-apart product.

Polaroid continued to improve the Land camera and in 1965 introduced the low-priced Swinger that was an immediate success. In 1972, after almost a decade of research, the SX-70 system became a reality. This was a sophisticated 26-ounce camera that used an integral film containing 11 coated layers of chemical compounds between two plastic bases and built into a single sheet. This sheet would eject from the camera immediately after exposure and develop into a color image in daylight and required no peeling. The SX-70 eventually became a best seller.

Polaroid had no competition in the instant photographic field until April 1976 when Kodak introduced to the market an integral system consisting of its PR-10 film and EK-4 and EK-6 cameras. A 14-year legal battle then ensued between Polaroid and Kodak. The Massachusetts Federal District Court handed down a decision on September 13, 1985, that ruled Kodak infringed on several Polaroid patents. As a result, Kodak was ordered to stop manufacturing and selling PR-10 film and EK-4 and EK-6 cameras. A federal judge on October 12, 1990, ordered Kodak to pay $909.4 million in compensation to Polaroid.

Land retired from Polaroid in August 1982 but continued to conduct research in several scientific fields at the Rowland Institute for Science at Cambridge, a research organization he created and financed in 1980. He was active in the scientific world until his death on March 1, 1991, in Cambridge, Massachusetts.

Land was honored many times. He collected 14 honorary degrees and 37 medals and awards, including the Presidential Medal of Freedom. He was inducted into the National Inventors Hall of Fame and was granted more than 533 patents. He was appointed to the faculties of Harvard and MIT, to the President's Science Advisory Committee, and to the President's Foreign Intelligence Advisory Board.

Internal Combustion Engine

The modern four-cycle internal combustion gas engine was the result of work done by Nikolaus August Otto who received the first U.S. patent for a commercially successful four-cycle internal combustion gas engine. He was awarded No. 194,047 on August 14, 1877.

Like so many other great inventions, the efforts of many people contributed to the final culmination of the internal combustion engine. It all began with Abbe Hautefeuille who, in 1678, proposed to raise water by the explosion of gunpowder. The gunpowder was exploded in a closed container, and a partial vacuum, formed by cooling the products of combustion, sucked up and raised the water from a reservoir by atmospheric pressure. In 1680, Christian Huyghens and his assistant, Denis Papin, of Holland constructed a gunpowder engine and were the first to use a cylinder and piston in a heat engine. John Barber of England was granted British Patent No. 1,833 of 1791 for an engine that utilized coal gas, derived from the distillation of coal, as fuel where the combustion of the gas was performed at constant pressure.

In 1794, Robert Street, an American, received British Patent No. 1,983 for the first workable internal combustion engine with a cylinder and piston. This engine was not self-acting, since the piston had to be lifted by an attendant for every stroke. It was, however, the first to use vaporized petroleum products as fuel. In 1799, French engineer Phillippe Lebon constructed a coal gas engine that compressed a mixture of gas and air before iginition. Unfortunately, Lebon was assassinated in 1804 before the details of his invention were completely worked out. In 1823, Samuel Brown of England invented the first gas engine that performed actual industrial work. This was the first internal combustion engine of the vacuum type, and for this, Brown received British Patent No. 4,847 of 1823. Between 1827 and 1832, he made engines that drove pumps, a boat, and a carriage. Samuel Morey of Oxford, New Hampshire, received on April 1, 1826, a United States patent for a gas or vapor engine. His engine had two cylinders, 180-degree cranks, poppet valves, a carburetor, an electric spark, and a water cooling device. For fuel, a mixture of turpentine and air was used.

In 1833, W. L. Wright of England patented a vertical double-acting gas engine, in which a mixture of coal gas and air was introduced on each side of the

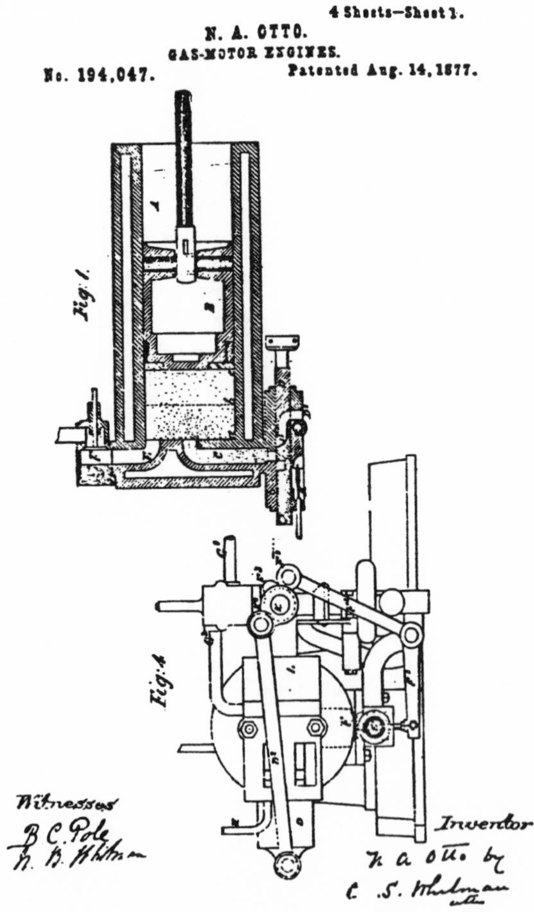

N. A. Otto's concept for Gas-Motor Engines was patented on August 14, 1877

piston. This engine is described in British Patent No. 6,525 of 1833. William Barnett of Brighton, Sussex, England, received British Patent No. 7,615 of 1838 for an engine in which the compression of 25 pounds per square inch of both air and gas was performed by separate pumps. Stuart Perry of New York City was awarded U.S. Patent No. 3,597 on May 25, 1844, for a gas engine that used liquid hydrocarbons as fuel with the charge under pressure at the time of ignition. This engine also had a carburizing device. In 1855, Alfred Drake of Philadelphia, Pennsylvania, was granted U.S. Patent No. 12,715 for an ignition engine that had a 16-inch diameter cylinder, a piston connected to a crank-shaft on which was a heavy flywheel, all surrounded and cooled by a water jacket. Eugene Barsanti and Felix Matteucci of Italy were granted British Patent No. 1,655 of 1857 for an atmospheric engine having a free piston and an auxiliary counter-piston that drew in a charge of gas and drove out the combustion products.

In 1860, Jean Joseph Etienne Lenoir of France built a one and one-half horsepower gas engine and placed it in a road vehicle. This was a horizontal, double-acting, single cylinder engine with slide valves. There was no compression, and the mixture of gas and air was ignited at mid-stroke by an electric spark-plug. This engine enjoyed some success with several hundred being built in France and England. A Frenchman by the name of Alphonse-Eugene Beau de Rochas in 1862 proposed the principle of a four-stroke cycle gas engine, but the principle existed in theory only since Beau de Rochas was a theoretician rather than a practical engineer. He never built an engine himself, though he did take out several patents. In 1865, Pierre Hugon of France invented a double-acting, vertical-type, gas engine that used a flame for ignition. In 1873, Siegfried Marcus of Austria built a four-stroke gas engine but abandoned it. In 1874, George Bailey Brayton of Boston, Massachusetts, invented a two-cycle gas engine that compressed the gas and air mixture before explosion and operated on liquid petroleum. These engines were produced in large numbers in the United States and were quite successful before the introduction of the four-cycle engine. However, it was Nikolaus August Otto in 1876 who built and perfected the first successful internal combustion engine using the four-stroke cycle. He became known as the "father of the internal combustion engine."

Nikolaus August Otto was born June 14, 1832, at Holzhausen on the Heather in Nassau, Germany. His father was an innkeeper at Holzhausen. Young Otto completed school with high marks and served a three-year apprenticeship with a neighborhood merchant. He then moved to Cologne and became a salesman. In 1860, he became aware of the Lenoir gas engine and thought it might be a promising line for an energetic salesman. In 1861, he began to experiment with these gas engines and built a simple two-stroke engine. In 1864, he met Eugen Langen, an engineer with financial resources, and together they founded the N. A. Otto Company. In 1866, he patented a vertical atmospheric engine that won the gold medal at the Paris Exhibition in 1867. The Otto model was more than twice as economical on fuel consumption as the Lenoir engine.

In 1871, the partners founded the firm Gasmotorenfabrik Deuz near Cologne at Deutz. The following year they employed Gottlieb Daimler, who later built the Mercedes automobile and named it after his daughter. They also employed Wilhelm Maybach and Franz Reuleaux to run the plant while Otto and Langen turned their attention to making a practical four-stroke engine. In 1876, Otto built the first successful four-stroke engine, calling it the Otto "silent" gas engine. The next year, patents were secured in Germany and in many other countries, including the United States. It was a sensational success at the Paris Exhibition in 1878. Over the next 17 years, nearly 50,000 Otto engines were built and sold.

The company had no sooner achieved success, however, before it faced litigation. Beau de Rochas entered into litigation to uphold his own patent rights, and in 1886 the German High Court ruled against Otto's patent. However, the damage was more to Otto's pride, than to his fortune. The Gasmotorenfabrik continued to pay Otto good dividends even after losing the patent suit. In the 18 years he was with the firm, Otto received in the form of salary, royalties, and dividends a total of 3.3 million marks.

The long 10-year court ordeal, in which Otto legally represented the firm, sapped his strength, and he required frequent rest cures to prevent personal collapse. He never completely recovered, and on January 22, 1891, Nikolaus August Otto died peacefully in his sleep at his Cologne home.

Jet Engine

The use of jet-propelled aircraft as a means of fast travel is due to the efforts of many people. No one, however, contributed more than Frank Whittle, who received the first U.S. patent for a jet engine for aircraft propulsion. The patent number was 2,404,334, issued July 16, 1946.

The history of jet propulsion dates back to 100 B.C. when Hero of Alexandria, Egypt, built a steam jet engine called an aeoliphile. In 1629, Giovanni Branca, an Italian engineer, built a steam turbine that applied the jet principle to operate primitive machinery. In 1678, Ferdinand Verbiest, a Jesuit in China, built a carriage powered by a jet of steam. In 1687, Sir Isaac Newton of England produced a model of a horseless carriage powered by jet propulsion and also stated the law of action and reaction. In 1781, James Ramsey of America built a steamboat using the jet principle for power. Ten years later, John Barber of England received British Patent No. 1,833 of 1791, the first patent for a gas turbine. In 1906, Rene Armengand and Charles Lemale of France built a continuous combustion turbine of 400-horsepower. In 1908, French engineer Rene Lorin received French Patent No. 390,256 for a ramjet engine using a reciprocating engine and exhaust jet. In 1923, Edgar Buckingham, an American engineer, presented a study of jet propulsion for aircraft. In 1926, A. A. Griffith of England designed a compressor to make the gas turbine a practical engine for driving a propeller type aircraft. However, it was Englishman Frank Whittle who conceived the idea of combining a gas turbine with jet propulsion for powering an aircraft.

Frank Whittle was born in Coventry, Warwickshire, England, on June 1, 1907. His father was a mechanic who owned and operated a small valve and piston-ring factory, and who had made several mechanical inventions. Young Whittle helped in his father's small factory, and it was there that the boy first learned about engines and how they worked. He attended public school in Coventry and entered Leamington College on a scholarship at age 11. He tried to enlist in the Royal Air Force (R.A.F.) in 1922 but was turned down because he was three inches too short. After months of stretching exercises, he was accepted by the R.A.F. in 1923 at the age of 16 as an apprentice. Later, he qualified as a pilot at the Royal Air Force College at Cranwell.

In 1928, Whittle was assigned to a fighter squadron as a test pilot at the Marine Aircraft Experimental Establishment. During this time, he wrote a thesis on

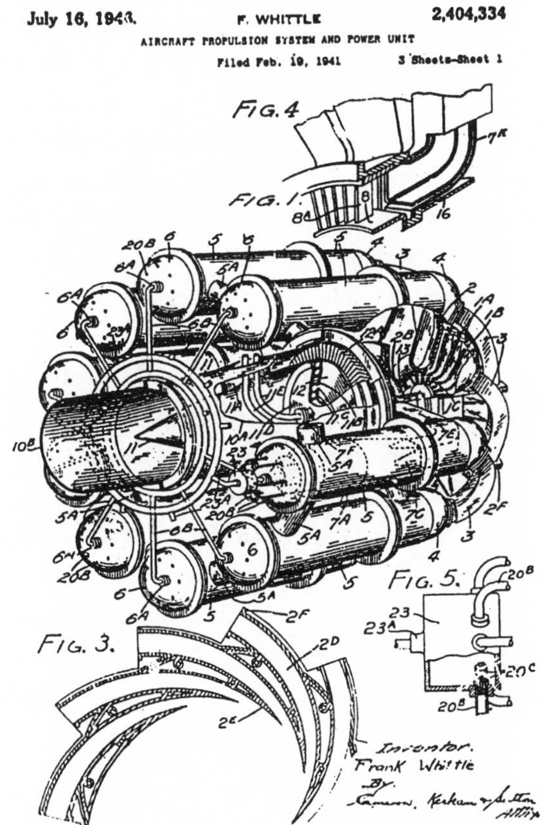

The patent for a jet engine for aircraft propulsion, issued July 16, 1946

the development of aircraft design showing that jet propulsion and a gas turbine were possible for driving a propeller. In 1929, he continued his study of propulsion at the Central Flying School at Wittering. It was there that he combined a gas turbine with jet propulsion called a turbojet engine, and received British Patent No. 347,206 in January 1930. Whittle took the sketches of the new airplane and its engine to the British Air Ministry, a department of the British government that decided which airplanes were to be built. The Air Ministry turned down the idea, saying it was impractical. Whittle then tried to interest various private firms in 1930 but without success. Over the next several years, his fellow R.A.F. officers called Whittle's idea for an engine, Whittle's Flaming Torch Hole. He did not renew his patent in 1935 because of his many disappointments and because he could not afford the five pounds for the patent renewal.

In May 1935, while Whittle was at the University of Cambridge studying mechanical sciences, he was approached by two former R.A.F. officers, R. D. Williams and J. C. B. Tinling. These men suggested that Whittle renew his patent as well as his interest in jet propulsion. The inventor agreed and with the help of an investment banking firm, O. T. Falk and Partners, formed Power Jets Ltd. in March 1936. Some of the shares allotted to Whittle were held in trust for the British Air Ministry. In 1937, the R.A.F. placed Whittle on Special Duty List, enabling him to devote full time to the development of the engine. The British Thomson-Houston Company at Rugby was awarded a contract to design and build an engine according to Whittle's requirement, and the Scottish firm of Laidlaw, Drew & Company was awarded a contract for the development of the combustion system. On April 12, 1937, the first jet engine designed and constructed by Whittle and his associates was tested under its own power.

In 1939, the Air Ministry changed its mind and placed a contract with Power Jets Ltd. for a flight engine. The first British flight powered by the Whittle W.1 engine was on May 14, 1941, when the Gloster aircraft E 28/39, unofficially called the *Pioneer,* took to the air with Flight Lieutenant Phillip E. G. (Jerry) Sayer as pilot.

In the United States, the first jet flight was on October 1, 1942, when the Bell XP-59A Airacomet aircraft, fitted with two General Electric Company Whittle type jet engines, reached a top speed of more than 500 m.p.h. with Robert M. Stanley at the controls.

The Germans were also experimenting with jet-propelled aircraft at the same time as Whittle. Hans von Ohain, a student of applied physics and aerodynamics at the University of Goettingen, had patented a centrifugal-type turbojet engine in 1934. However, it remained for Frank Whittle to design and build a jet engine that would eventually receive worldwide attention.

Many honors came to Whittle after he successfully completed his turbojet engine. He retired from the R.A.F. in 1948 with the rank of air commodore and was knighted the same year by King George VI, as Sir Frank Whittle. Shortly thereafter, the British government granted him a tax-free gift of 100,000 pounds for his pioneering work on jet propulsion. In 1953, his book *Jet, The Story of a Pioneer* was published. In 1977, he became research professor at the U.S. Naval Academy in Annapolis, Maryland. In 1987, Sir Frank Whittle and Hans von Ohain were both awarded the National Air and Space Museum Trophy.

Laser

Perhaps one of the most important inventions of the 20th century is the invention of the laser. From the operating room to the battlefield, its uses are too numerous to list. The first U.S. patent issued for the laser was awarded to Charles Hard Townes and Arthur Leonard Schawlow. The number was 2,929,922, issued on March 22, 1960.

The history of the laser is really the story of four men: Charles Townes, Arthur Schawlow, Theodore Maiman and Gordon Gould. However, these men in developing the laser built on the foundation of other scientists. These include Christian Huyghens, a Dutch physicist and inventor of the pendulum clock, who in 1678 suggested that light consists of tiny waves; Scottish physicist James Clerk-Maxwell who in 1850 discovered that electromagnetic radiation travels in waves at the speed of light; German physicist Max Planck who in 1900 formulated the quantum theory and showed that light travels in distinct chunks, called quanta, and comes in precise parcels of energy, called photons; Niels Bohr, a Danish physicist, and Ernest Rutherford, an English physicist, who in 1911 figured out that electrons travel around the nucleus in distinct orbits or energy levels and that if an electron drops from a higher orbit to a lower, it emits a bundle of energy; American physicist Albert Einstein who in 1916 predicted that electrons could be stimulated to emit light of a particular wavelength if the stimulus was additional light of that wavelength; English physicist Paul Dirac who in 1927 discovered that the two protons produced by stimulated emission are coherent; and Russian physicist V. A. Fabrikant who in 1951 proposed an amplification of stimulated emission as did Joseph Weber, an American physicist, in 1953 and Nikolai Basov and Aleksander Prokhorov, Russian physicists, in 1954. It remained for Townes, Schawlow, Maiman, and Gould to combine these teachings to produce a practical laser.

Charles Hard Townes was born July 28, 1915, in Greenville, South Carolina. His father, Henry Keith Townes, was an attorney. Young Charles attended elementary school in Greenville and graduated from the local high school in 1931. He secured a fellowship from Furman University in Greenville and graduated with BA and MA degrees in 1935. He studied for a master's degree in physics at Duke University, graduating in 1937. Two years later, he earned a PhD degree in physics in from the California Institute of Technology. His dissertation, under the direction of Professor W. R. Smythe, was published in 1939 in *Physical Review*.

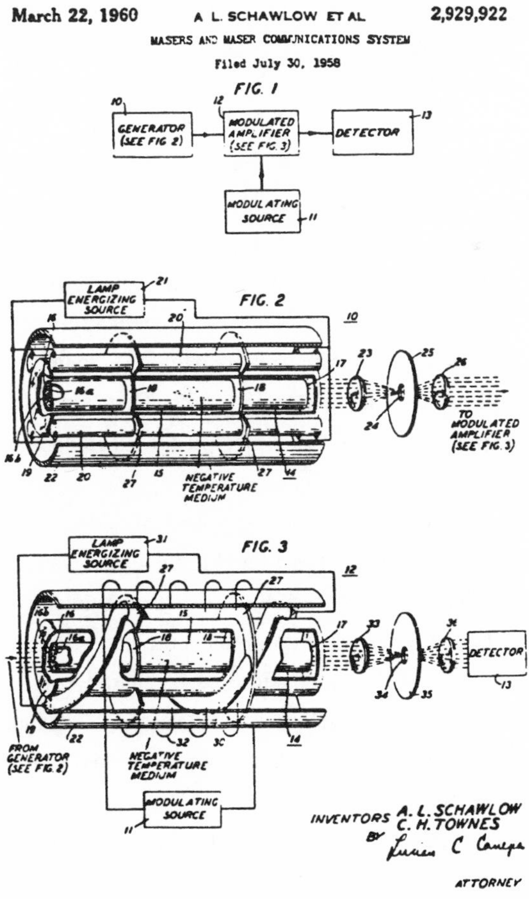

The laser, a 20th century invention, issued on March 22, 1960

Townes spent World War II and some years thereafter working on the design of radar bombing systems at the Bell Telephone Laboratories in New York City and in Murray Hill, New Jersey. In 1948, he accepted a position as associate professor of physics and director of the radiation laboratory at Columbia University, supervising research using microwaves.

On April 26, 1951, Townes and Arthur Schawlow attended a meeting of physicists in Washington, D.C. It was there on a park bench in Washington's Franklin Park that Townes suddenly realized the conditions needed for the amplification of the stimulated emission of microwaves. He formulated the problem as a thesis topic on the back of an old envelope and turned it over to two of his graduate students at Columbia University, James P. Gordon and Herbert Zeiger. Three years later, Townes, Gordon, and Zeiger produced the first maser, an acronym for microwave amplification by stimulated emission of radiation.

In September 1957, Townes sketched a design for a device called an "optical maser" that would emit infrared or visible light. He and his brother-in-law Arthur Schawlow developed the laser, an acronym for light amplification by stimulated emission of radiation. The two men filed a patent application on July 30, 1958, which matured into U.S. Patent No. 2,929,922—the first laser patent. They also published a paper on the subject in the December 15, 1958, issue of *Physical Review.*

Many honors were bestowed on Townes, including those by the Franklin Institute, National Academy of Sciences, and American Academy of Arts and Sciences, with perhaps the greatest being the Nobel Prize in physics that he shared with the Russian scientists, Nikolai G. Basov and Aleksander M. Prokhorov, who independently worked out the laser theory.

Arthur Leonard Schawlow was co-inventor of the laser with Charles Townes. Schawlow was born in Mount Vernon, New York, on May 5, 1921. His father, an insurance agent, had emigrated to New York from Riga, Latvia, some 10 years before. When young Schawlow was three years old, the family moved to Toronto, Canada. There, he attended the Winchester Elementary School and graduated from high school in 1937 at the Vaughan Road Collegiate Institute.

Schawlow won a scholarship in mathematics and physics and received his BS in 1941 from the University of Toronto. He taught classes for military personnel at the university and received his PhD in physics in 1949 from the University of Toronto, his thesis research being optical spectroscopy.

He then spent two years at Columbia University on a postgraduate fellowship. It was there that he met Charles Townes and worked with him on microwave spectroscopy research. It was also there that he met and married Townes's youngest sister Aurelia. In 1951, Schawlow accepted a position at Bell Telephone Laboratories in Murray Hill, New Jersey but continued his association with Townes. They co-authored a book, *Microwave Spectroscopy*, published in 1955.

During the next few years, Schawlow and Townes continued working together, mostly on weekends, on the maser, which Townes and two colleagues, James P. Gordon and Herbert Zeiger, had developed in December of 1953. An outgrowth of this work was the filing of a patent application on July 30, 1958. This resulted in the issuance of the first laser patent on March 22, 1960, and the publication of an article entitled, "Infrared and Optical Masers," in *Physical Review* on December 15, 1958.

In 1960, Schawlow returned as a visiting professor to Columbia University. In 1961, he became professor of physics at Stanford University. Since 1978, he has been the J. G. Jackson–C. J. Wood Professor of Physics at Stanford.

Many honors have been awarded to Schawlow. In 1976, he received the Frederick Ives Medal of the Optical Society of America; in 1962, the Stuart Ballantine Medal of the Franklin Institute; in 1963, the Thomas Young Medal and Prize of the Institute of Physics in London; and in 1981, the Nobel Prize for physics that was shared with Nicolaas Bloembergen, a Harvard University physicist, "for their contribution to the development of laser spectroscopy."

The first person to successfully demonstrate the laser was Theodore Harold Maiman. He was born in Los Angeles, California, on July 11, 1927. His father was an electrical engineer. Young Maiman worked his way through the University of Colorado by repairing electrical appliances and graduated in 1949. He earned his PhD in 1955 at Stanford.

He secured a job with the Hughes Aircraft Company's Research Laboratories in Malibu, California, and became interested in Townes's maser. He designed a ruby cylinder with its ends carefully polished flat and parallel and covered with a thin silver reflective coating, leaving a small hole for the exit of radiation. In May 1960, after being fed with light from a xenon flash lamp, a monochromatic and coherent flash of light was emitted. On July 7, 1960, he announced to the press the first demonstration of a laser.

Maiman offered to publish his laser experiments in *Physical Review Letters*, but the editors rejected his paper. The British journal *Nature* did publish his 300-word article in 1960. Commercial laser manufacture soon began, and one of the first companies in the business was founded in 1962 by Maiman in Santa Monica, California, and called Korad Inc. Maiman was granted a patent on the ruby laser and eventually sold his interest in Korad Inc. to the Union Carbide Corporation. He is now vice president for technology at TRW Inc.

One of the most interesting individuals involved in the development of the laser was Gordon Gould. He was born in Pittsburgh, Pennsylvania, in 1920. His father was the editor of *Scholastic Magazine*. After an early education in Pittsburgh, young Gould received degrees in physics from Union College and Yale University. During World War II, he worked on the secret Manhattan Project that developed the A-bomb. In the mid 1950s, he began working toward a PhD in microwave spectroscopy at Columbia University under the direction of Nobel laureate Polykarp Kusch. The director of Columbia's Radiation Laboratory at that time was Charles Townes, inventor in 1951 of the maser.

One night in October 1957, Gould, unable to sleep, conceived an idea and wrote it down on 15 pages of a notebook, with the title: "LASER: Light Amplification by Stimulated Emission of Radiation." He had his notebook notarized by a Bronx candy store owner, Jack Gould, who also served as a notary public. These events preceded by more than a year the publication of a paper in *Physical Review* (December 1958) on the subject by Townes and Schawlow. Gould, on the advice of a patent attorney, thought erroneously that he needed a working model of a laser before he could file a proper patent application. So he waited until April 6, 1959, to apply. He later discovered that Townes and Schawlow had filed on July 30, 1958.

To earn money to build a working model, Gould dropped out of Columbia University and joined a small research firm in Syosset, New York, called TRG Inc. This company applied for a $300,000 military contract with the Department of Defense. Because the government liked the idea of the laser as a military weapon, it gave TRG Inc. $1 million instead of the requested $300,000. The government also decreed the project classified which meant Gould was a security risk because he had attended sessions of a Marxist study group some 16 years earlier. Consequently, Gould couldn't even look at his own notebooks.

Gould had filed his patent application on April 6, 1959. The Patent and Trademark Office required his application to be divided into five separate and distinct inventions. These divisional applications became involved in five costly and time-consuming interference proceedings. He lost three of these interferences but won two. In 1970, after spending $250,000 prosecuting the rights to the laser patent, TRG Inc. was acquired by Control Data, and Gould reacquired his patent rights. He spent another $100,000 of his own money in legal fees.

In 1974, unable to personally finance any more legal battles, Gould sold part of his interest in his patent position to the Refac Technology Development Corporation, a New York patent-licensing company headed by Eugene Lang. In return, Refac agreed to press for patents and patent licenses once patents were issued. Refac prosecuted Gould's applications before the Patent and Trademark Office and various courts and got results. On October 11, 1977, over 16 years after he had applied, United States Patent No. 4,053,845 was issued to Gould on an optically pumped light amplifier. On July 17, 1979, the Patent and Trademark Office issued to Gould United States Patent No. 4,161,436 for a method of energizing a material utilizing light amplifier apparatus. When the optically-pumped patent was issued, Refac immediately demanded royalties of about 5 percent from laser manufacturers. Some refused, and the first of many infringement suits was brought against the Control Laser Corporation of Orlando, Florida. This is perhaps one of the most complex patent cases in the history of the United States. The 18 pages of patent disclosure are accompanied by a 500-page "file wrapper" detailing the patent's legal history.

In 1979, Gould sold part of his remaining interest to Gary Erlbaum, who owned a money-losing chain of retail building-supply stores called Panelrama Corporation, for $300,000 in cash and notes worth $2 million. Erlbaum liquidated Panelrama and changed its name to the Patlex Corporation.

In addition to the infringement litigations, Gould encountered other problems with these issued patents. On July 1, 1981, Public Law No. 96-517 went into effect. This law allows a person to call to the attention of the Patent and Trademark Office any prior art that may have a bearing on the patentability of a claim in any unexpired patent issued before, as well as after, that date. In 1982, Control Laser and Bell Laboratories, and in 1983 General Motors, requested the Patent and Trademark Office to reexamine the 4,053,845 patent. In 1982, Lumonics Research and General Motors requested the Patent and Trademark Office to re-examine the 4,161,436 patent. These issues, as well as the infringement issues, have not been completely settled. However, if the patents are upheld, Patlex, Gould, Refac, and Gould's patent law firm of Lerner, David, Littenburg & Samuel who handled the case in return for an interest in the patents, will share royalties that could run into tens or even hundreds of millions of dollars over the life of the patents.

Gould is now vice president of Optelecom Inc., a fiber-optic communication company in Gaithersburg, Maryland. He recently received the Inventor of the Year Award presented by the Association for the Advancement of Invention and Innovation.

Perhaps one day history will give credit to the first to invent the laser. For now, however, the first U.S. patent for the laser belongs to Charles Townes and Arthur Schawlow.

Lathe

During the early years of the United States, there arose a need for wood to be shaped in irregular forms. These forms included gunstocks, axe handles, and shoe lasts, to name a few. The man responsible for inventing a machine to do this was Thomas Blanchard, who received the first U.S. patent for a lathe that would make these irregular wooden materials. He was awarded the patent on September 6, 1819. This patent was extended twice by Congress in 1834 and 1848 for a total period of 42 years.

The lathe is one of the oldest machine tools. Hand-powered lathes were widely known in the Mediterranean area by the 5th century B.C. Later, the lathes were foot-powered where a forward and backward rotary motion was given to the workpiece by the alternative movements of a cord secured at one end to a foot treadle and the other end to a spring pole. In the 14th century, a continuous rotary motion was applied to the workpiece by the combination of an endless rope, crank, and flywheel. Around 1500, water-power became common as a means for turning lathes.

The first screw-cutting lathe was invented in 1569 by the French mathematician, Jacques Besson of Besanger, France. In 1710, Christopher Polhem of England built a lathe large enough to fashion articles needed in industry. In 1794, Joseph Bramah, an English machinist, invented the slide-rest lathe. In 1800, Henry Maudslay, also of England, designed the first precision industrial lathe that had an all-metal frame, a mechanical tool holder, and a precision lead screw to drive the tool holder across the workpiece. The Maudslay lathes were not offered for sale, but one of his students, Sir Joseph Whitworth of Manchester, England, started manufacturing industrial machine tools for sale, including the lathe, in 1833. On September 6, 1819, Thomas Blanchard of the United States received the first U.S. patent for a lathe for making irregular items.

Thomas Blanchard was born June 24, 1788, in Sutton, Worcester County, Massachusetts. His father, Samuel Blanchard, was a farmer. Young Blanchard stammered badly which brought him so much ridicule that school became unappealing to him. He soon quit to help on his father's farm and in his spare time took an interest in woodworking and metalworking with a penknife and chisel. At age 13, he invented an apple parer that could easily peel more fruit than six hand-peelers.

His elder brother Stephen operated a tack-making factory at West Milbury, Massachusetts. Thomas went to work for his brother making tacks one by one by hand with a hammer and vice. Soon after he

No Printed Specification Available

T. BLANCHARD.

Turning Irregular Forms.

Patented Sept. 6, 1819.

Date altered by act of Congress to Jan. 20, 1820.

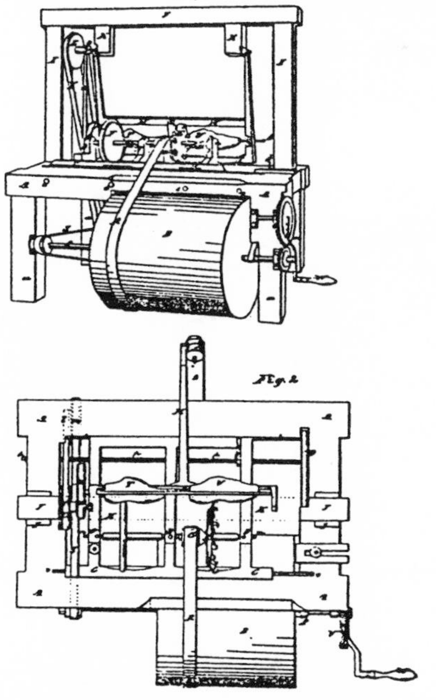

This lathe, designed for the crafting of irregular wooden materials, was patented on September 6, 1819 by Thomas Blanchard

arrived, he invented a counter with a bell that rang after 100 tacks were counted. The tacks were then placed in a carton for sale. Blanchard was 18 when he began to build a tack-making machine. After six years of work at odd times, his machine could automatically turn out 500 tacks a minute. He sold the patent rights to this invention for $500.

Blanchard's success with his tack-making machine brought him fame throughout New England. His work attracted the attention of a proprietor of an armory in Springfield, Massachusetts, that produced muskets for the army. He employed Blanchard to build a lathe that would turn the barrel and produce the flat and oval portions of the breech in one operation. In less than a week, he had built such a lathe. He was next asked to produce a lathe that would turn gunstocks. Within a month, Blanchard built a lathe that carved a gunstock so perfectly that sandpaper was hardly needed. It worked equally well in producing other articles of irregular shape, such as shoe lasts, hat blocks, wheel spokes, and axe handles. On September 6, 1819, he received a U.S. patent for this lathe, which consisted of a friction-wheel that followed the contour of a pattern and a cutting-wheel secured to the same shaft that cut the workpiece into an exact replica of the pattern.

Blanchard continued to work for the United States Armory at Springfield. In 1822, he built a large copying lathe that could carve two gunstocks in an hour. This lathe is still on view in the Museum of the United States Armory at Springfield. The United States Armories at Springfield and Harper's Ferry paid him a royalty of nine cents for each gunstock turned on his lathe. The English government ordered 10 of Blanchard's lathes for making gunstocks for its army, paying him $40,000 for them. When his lathe patent expired in 1833, he petitioned Congress for an extension that was granted in 1834 and again in 1848. His patent was in force for a total of 42 years.

Blanchard also made other inventions. He invented a clamping machine where large timbers could be bent to desired forms without breaking. He received $150,000 for its application to ship timbers. He also invented a machine for cutting and folding envelopes. In 1825, he built and patented a steam carriage. He also designed and built several kinds of shallow-draft steamers.

Blanchard moved to Boston during the latter part of his life, where much of his time was devoted to acting as an expert in patent cases. He died in Boston on April 16, 1864.

Linotype

The man most responsible for lifting the art of printing out of the Middle Ages was Ottmar Mergenthaler, who received U.S. Patent No. 304,272 on August 26, 1884, for a Linotype machine. This was the first U.S. patent for this type of machine.

In the 1450s, the first book, the *Gutenberg Bible*, was printed from movable metal type set by hand. In 1822, Dr. William Church, an American living in England, invented the first typesetting machine and received the first patent for a mechanical typesetter. In 1873 and for the next 20 years, there existed the most complicated typesetting machine ever built. It was created by James W. Paige of Hartford, Connecticut, weighed more than three tons, and contained over 18,000 parts. Only two were built at a cost of a million dollars and neither was operative for commercial demands. Mark Twain lost a considerable amount of money in backing this venture. In 1874, Charles Kastenbein, a German living in England, developed the first typesetting machine to print a newspaper, the London *Times*. Then in 1884, Ottmar Mergenthaler revolutionized the printing world with his Linotype machine.

Ottmar Mergenthaler was born in Hachtel, Germany, on May 11, 1854. His father was a schoolmaster in Wurttemberg. Young Mergenthaler received an ordinary grade-school education. At the age of 14, he began an apprenticeship in watch and clock-making with a distant relative named Hahl, in nearby Bietigheim. Hahl had a son, August Hahl, who had gone to the United States and set up a scientific instrument shop in Washington, D.C., where he made signal apparatus for the U.S. government. In 1872, Mergenthaler also went to the United States to work in the instrument shop of Hahl & Company. Shortly thereafter, the company and Mergenthaler moved their operation to 13 Mercer Street, Baltimore, Maryland.

The official reporter for the United States Senate at this time was James O. Clephane who had the arduous task of copying by hand the volume of words spoken by the legislators. He wished for a machine that would put the Senate records in printed form. Clephane and a group of friends became interested and financially backed a writing-machine for printing words on lithographic stone. This machine was invented by a West Virginian, Charles G. Moore. The device turned out to be unsatisfactory for

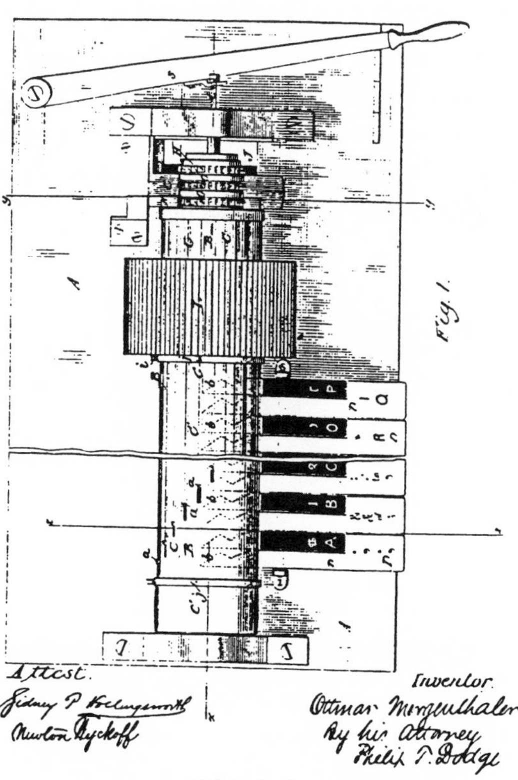

Ottmar Mergenthaler's "Matrix Making Machine," the Linotype, patented August 26, 1884

Clephane's work, so in 1876 Moore took the machine to August Hahl's machine shop in Baltimore for some needed improvements. Mergenthaler was assigned the task and corrected the defects and, under orders from Clephane, built a full-sized machine in 1877. However, this machine also did not give satisfactory results, and the lithographic project was abandoned.

Mergenthaler built another machine, at Clephane's suggestion, that impressed characters into a strip of papier-mache to make a mold into which type metal could be poured. This machine did not work successfully, because the papier-mache strips clung to the cast metal. Consequently, this project was also abandoned in 1879.

Out of these failures, however, came Mergenthaler's great idea in 1883. During the next two years he built a machine in which separate, single-letter female metal matrices were caused to be set in a line by the action of the keyboard. Metal was then poured in to make a solid printing line. When the type metal hardened, the line was placed in a tray called a galley for the press, and the matrices were separated and carried back mechanically to their proper compartment. The compositor then set the next line. These machines and the improvements that followed were completely successful.

The first demonstration of this machine was July 3, 1886, at the *New York Tribune*, whose owner, Whitelaw Reid, coined the term "Linotype" as he watched. Mergenthaler soon began making and selling his typecasting-machines. The *New York Tribune* bought 20, the *Louisville Courier-Journal* 20, the *Chicago News* 20, and the *Washington Post* 12. The newly formed Mergenthaler Printing Company sold approximately 200 of these machines.

A syndicate of newspaper men, assembled by Clephane, bought a controlling interest in the company for $300,000. These men were Whitelaw Reid of the *New York Tribune*, W. H. Haldeman of the *Louisville Courier-Journal*, Victor Lawson and M. E. Stone of the *Chicago News*, Henry Smith of the *Chicago Inter Ocean*, W. H. Rand of Rand, McNally & Company of Chicago, and Stilson Hutchins of the *Washington Post*. A change in company policy caused a break between the board of directors and Mergenthaler, resulting in his resignation in 1888. However, Mergenthaler independently continued to make improvements to his Linotype, resulting in more than 50 patents that he sold to the syndicate.

When his health declined, Mergenthaler retired from active work on the Linotype. He died October 28, 1899, in Baltimore after a five year battle with tuberculosis.

Before his death, he received many honors. He was naturalized on October 9, 1878. He was awarded a medal by Cooper Union of New York for his invention. The Franklin Institute of Philadelphia awarded him the John Scott Medal and the Elliott Cresson Medal.

Lock

The first commercially successful lock and key was the pin-tumbler cylinder lock, using a small flat key. It was invented by Linus Yale Jr. who received U.S. Patent No. 31,278 on January 29, 1861.

The invention of the lock probably stems from the time man first acquired goods he wanted to safeguard from others. The oldest surviving locks were found in the ruined palace of Khorsabad, built by the Assyrians about 8000 B.C. The Egyptians around 2000 B.C. were the first to invent a true key-operated lock made of wood. About 700 B.C., the Greeks were the first to use the keyhole in the lock itself. The Romans improved the Egyptian and Greek locks by incorporating springs, instead of depending on gravity, to press the pins into the bolt. They were also the first to use metal locks and the first to use portable or padlocks.

In 1784, Joseph Bramah of England, inventor of the hydraulic press, introduced the modern lock and key that became popular in Europe. This lock had several sliders and two barrels, the inner one shooting the bolt. In 1817, Jeremiah Chubb, a Portsmouth, England, ironmonger, invented the Chubb lock. This lock employed six control spring-operated levers that had to be raised to an exact height by the key. Other locks that found their way into general use were the mortised lock of Philos and Eli Blake in 1835, the Andrews permutation lock in 1841, and the Day and Newell parautoptic lock in 1843. However, it was Linus Yale Jr. on January 29, 1861 who received the first U.S. patent for a pin-tumbler cylinder lock that was commercially accepted in the United States and Europe.

Linus Yale Jr. was born at Salisbury, Herkimer County, New York, on April 4, 1821. His father, Linus Yale Sr. was a mechanic and an inventor. Little is known about young Yale's early life except that he was well educated and inherited his father's mechanical ability. In his early years, he studied art and for about 10 years struggled along as a portrait painter.

In 1840, Linus Yale Sr. invented and patented a bank lock, which he began to manufacture in Newport, New York. For many years Linus Jr. worked with his father in Newport. It was there that he invented his first bank locks and received U.S. Patent No. 8,071 on May 6, 1851, and U.S. Patent No. 9,850 on July 12, 1853. In 1855, he moved to Philadelphia, Pennsylvania, and set up his own locksmith shop.

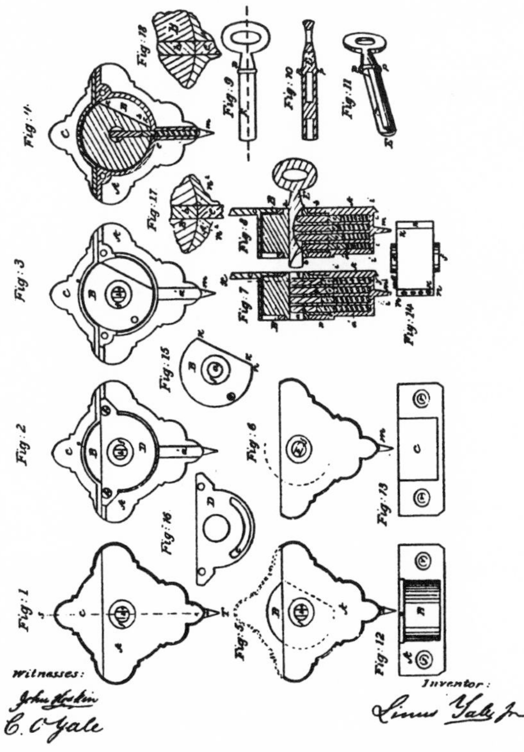

The pin-tumbler cylinder lock, invented by Linus Yale Jr. and patented January 29, 1861

There, he met John Henry Towne, who later made the name Yale famous worldwide.

Yale had his most creative and productive years in Philadelphia. He received patents on a clamp for a joiner's square, a screw tap, a mechanic's vise, and an alarm clock. But the work for which he will be remembered is the invention of locks. He invented the Yale Infallible Bank Lock, a lock in which the key was made up of parts that could be separated and reassembled to change the combination. His next invention was the Yale Magic Bank Lock, an improvement on his first product. At the request of the government, he invented the Yale Double Treasury Bank Lock. Next came the Monitor Bank Lock, the first of the dial or combination bank locks. His next invention was the Yale Double Dial Bank Lock, which required two keys to open. These locks made him known throughout the country. The lock that made him famous was the pin-tumbler, cylinder lock with the thin flat key. He received a U.S. patent for this lock on January 29, 1861, and for an improved version on June 27, 1865.

Yale was not a businessman. His time was spent mostly on his inventions and consulting work on bank locks. In 1868, William Sellers introduced him to John Henry Towne, a wealthy foundry owner. In October of that year, Towne established the Yale Lock Manufacturing Company with his son Henry Robinson Towne and Yale as partners. They began the construction of a plant in Stamford, Connecticut. However, the partnership lasted less than two months. While on a business trip to New York City, Linus Yale Jr. died of a heart attack on December 25, 1868.

Henry Robinson Towne reorganized the company and became president of the Yale & Towne Lock Company. It was Towne's decision to stamp the name Yale on every lock and key that left the Stamford plant as a memorial to Linus Yale Jr.

Machine Gun

Although not fully automatic, the invention of the machine gun is generally ascribed to Richard Jordan Gatling, who received U.S. Patent No. 36,836 on November 4, 1862. However, the first U.S. patent (No. 15,315) for a machine gun was awarded to C. E. Barnes of Lowell, Massachusetts, on July 8, 1856, and is considered the forerunner of the many crank-operated weapons. This gun was ahead of its time, and the world had to wait for the Civil War and Gatling before the machine gun became a practical reality.

In the 14th century, gunpowder was invented. This precipitated the invention of the multi-barreled volley gun, also called the organ gun or the ribauld. This gun consisted of several barrels mounted parallel on a cart that could be fired like a machine gun by waving a lighted torch across the touch-holes. Leonardo da Vinci designed some of these guns. However, it was James Puckle, a London lawyer, received the first patent for a machine gun on May 15, 1718. The interesting fact about this patent was that the drawing showed round bullets to be used against Christians and square ones to be used against infidels. In early 1862, an unknown inventor came out with the Union Repeating Gun, also known as the "Coffee Mill." This was probably the first machine gun to kill a man in battle. On May 31, 1862, Captain D. R. Williams of Covington, Kentucky, built a machine gun called "a repeating one-pounder," which was used in the Civil War. Richard Jordan Gatling, on November 4, 1862, revolutionized the machine gun industry with his crank-operated, patented, machine gun.

The evolution of the modern machine gun did not stop with Gatling. In 1863, James O. Whitcomb, of New York, patented a four-barreled rapid-fire gun to be fired electrically. In 1885, Hiram Stevens Maxim of the United States and England patented the first truly automatic machine gun using the recoil principle. In 1892, John Moses Browning of the United States patented the first gas-operated machine gun. This was followed in 1896 by Laurence V. Benet and Henri A. Mercie of the United States who patented the Hotchkiss gun, named in honor of their company's deceased founder, Benjamin Berkeley Hotchkiss.

Richard Jordan Gatling was born September 12, 1818, in Hertford County, North Carolina. His father, Jordan Gatling, was a planter who invented a cottonseed sowing machine and a machine for thinning cotton plants. After completing all the available schooling, young Gatling began to teach school in

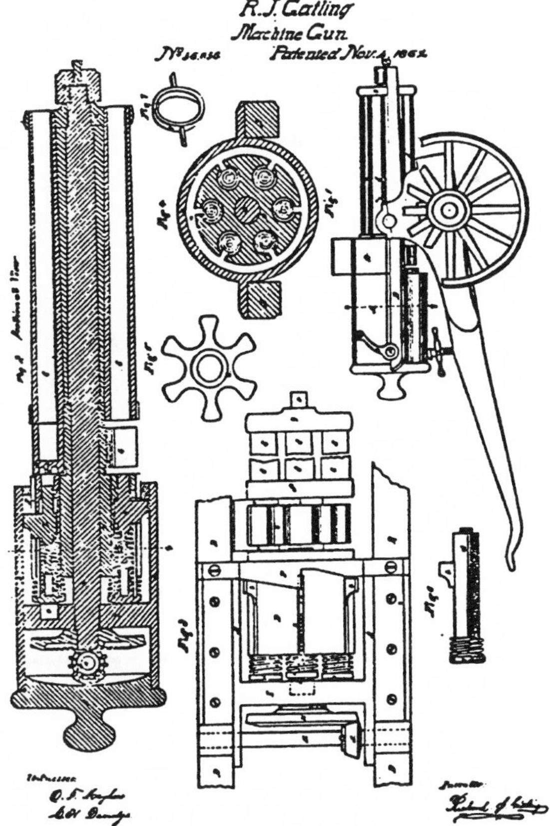

Richard Jordan Gatling's Machine Gun, patented November 4, 1862

Hertford County at age 19, but quit after a year to operate a country store. While still in his teens, he assisted his father in the construction of his machines and patented a rice planter in his own name at age 21. The previous year he had invented a screw propeller but did not patent it since he found that John Ericsson had anticipated him.

In 1844, Gatling went to St. Louis, Missouri, to have his rice-sowing machine manufactured, as well as a wheat-sowing machine that operated on the same principle. In the winter of 1845, while on a river-steamer from Cincinnati, Ohio, to Pittsburgh, Pennsylvania, he contracted smallpox. Because the river-steamer became ice-bound, he received no medical attention for two weeks. However, he did recover and decided to attend medical school so he could take care of himself and his family. From 1847 to 1848, he studied medicine at Laporte, Indiana, and in 1850 enrolled in the Medical College of Ohio at Cincinnati, where he earned a degree. Although there is no record of his ever practicing medicine, Gatling was always addressed as "Doctor."

His agriculture equipment business prospered so much that he opened plants in Springfield, Ohio, and Indianapolis, Indiana, in addition to the one in St. Louis. He also added a hemp-breaking machine and a steam-plow to his inventions and manufacture. The threat of Civil War caused him to turn his attention from agriculture equipment to weapons of war.

From a suggestion by Colonel R. A. Maxwell, Gatling drew up plans for a machine gun in 1861, which would later make him famous worldwide, and received United States Patent No. 36,836 on November 4, 1862. The gun had 10 barrels, which rotated about a central axis by turning a crank, and contained a hopper mounted over the breech. As each barrel reached the top, a cartridge from the hopper fell into its chamber. As the barrel reached the bottom of its orbit, it was fired and the empty cartridge case fell out.

The first gun was made in Indianapolis and had a firing capacity of 250 shots per minute. This success encouraged Gatling so much that he sought and received financial backing and moved to Cincinnati, Ohio, where Miles Greenwood & Company were contracted to make six weapons. Unfortunately, fire destroyed this factory and the nearly completed weapons, along with blueprints and patterns. Gatling was not easily discouraged, however, and again sought backing through McWhinny, Rindge & Company of Cincinnati. This company made 12 guns according to his 1862 patent.

Gatling continued to make improvements on his gun and was awarded a second patent on May 9, 1865. This improved gun was produced by the Cooper Fire-Arms Manufacturing Company of Philadelphia, Pennsylvania, after Gatling had ended his partnership with McWhinny, Rindge & Company. At first the government was reluctant to approve the gun, but on August 24, 1866, it was officially adopted for the United States Army. One hundred of these improved guns were made for Gatling and the army by the Colt Patent Fire-Arms Manufacturing Company of Hartford, Connecticut. For the next 30 years Gatling used the Colt Company to manufacture his gun. When he had improved his gun so it fired 1,200 rounds of ammunition per minute, he sold the patent rights to the Colt Company in 1870 and established his residence in Hartford.

The Gatling gun became the standard weapon in practically every foreign country. In 1876, the Gatling gun received the only award for machine guns at the International Exhibition in Philadelphia. Gatling received many honors both in the United States and abroad. He was president of the American Association of Inventors and Manufacturers for six years.

In 1900, he turned his attention again to agriculture equipment and invented a motor-driven plow. Before he had a chance to manufacture this plow, Gatling contracted an influenza-type disease and died February 26, 1903.

Magnetic Tape Recorder

Thomas Edison gave the world the first reproduction of sound, but magnetic recording was a giant step forward in reproducing sound and sight and made the technique of instant replay possible. The magnetic recorder was the invention of Valdemar Poulsen, who received the first United States patent for this device. It was issued November 13, 1900, and was assigned the number 661,619.

The first concept of magnetic recording was made by Oberlin Smith of Bridgeton, New Jersey. He filed with the United States Patent Office a caveat on a magnetic recorder. A caveat, which practice was abolished in the United States in 1909, was a notice to the Patent Office that an inventor had not yet completed the invention but was on the verge of doing so and needed further time to file an application. However, there is no record that he ever filed such an application and no evidence that he ever made a working recorder. Ten years later, Smith published his work on the recorder in the September 8, 1888, issue of *The Electrical World*, where he stated that he used fabric strips containing iron filings as the recording medium. However, on July 8, 1899, Valdemar Poulsen of Copenhagen, Denmark, filed a patent application with the United States Patent Office for a magnetic recorder. The patent was issued on November 13, 1900, and was the first United States patent ever issued for a magnetic recorder.

Poulsen was ahead of his time. The magnetic wire and tape recorder needed something to amplify the current to make it practical. The later invention of the vacuum tube by Sir John Fleming in 1904 made this possible. Even then the magnetic recorder had to wait for further development. In 1918, Kurt Stille, a German engineer, produced a practical magnetic wire recording system used in broadcasting. In 1927, J. A. O'Neill of the United States invented the first magnetic recording tape that used a paper base. In 1929, Pfleumer, an Austrian private researcher, developed a magnetic tape in which soft iron powder was mixed with an organic binding medium. Also in 1929, Louis Blattner of England made the first magnetic recorder with electronic amplification and steel tape called the Blattnerphone.

The great advance in magnetic recording came in 1935 when the German company Allgemeine Electrizitats Gesellschaft of Berlin developed the Magnetophon, the first system to use plastic tape

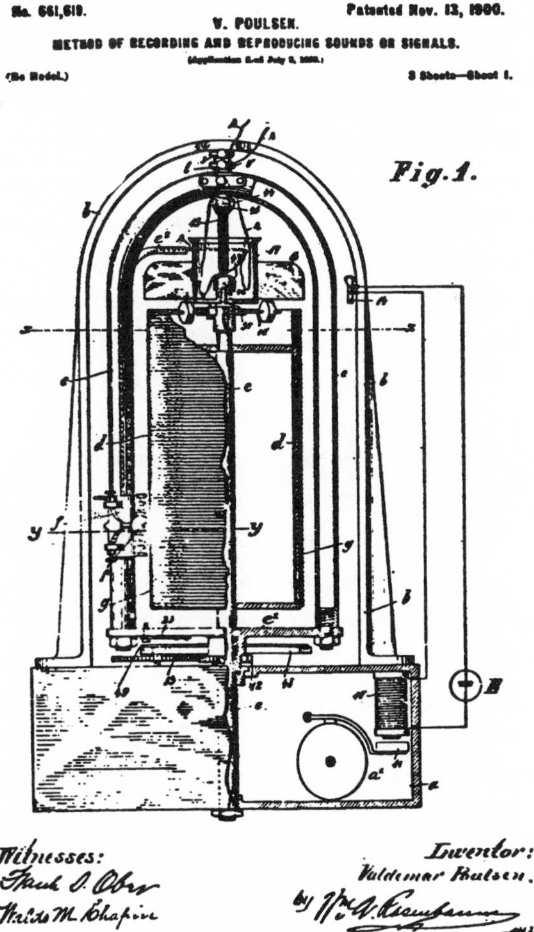

The magnetic recorder, invented by Valdemar Poulsen and issued November 13, 1900

coated with ferrous oxide. In 1940, improvement to this system was made by two Germans, H. J. Braunmuhl and W. Weber of the German State Broadcasting Service, when they applied a high-frequency a.c. bias to the oxide coated tape of the Magnetophon. During World War II, Magnetophons were widely used in Germany.

Little work was done in the United States on magnetic recording until 1937. In that year, S. J. Begun of the Brush Development Company of Cleveland, Ohio, produced the Soundmirror, and C. N. Hickman of the Bell Telephone Laboratories, produced the Mirrophone. In 1940, Marvin Camras of the Armour Research Foundation invented a much-improved wire recorder. He also made improvements in high-frequency bias, coating materials for magnetic tapes, and the use of magnetic sound in motion pictures.

After World War II, the United States Alien Property Custodian was awarded the patents to the German Magnetophon and its tape. The occupation authorities discovered 18 complete Magnetophons at the AEG plant in Berlin and portioned them out between the British, French, and U.S. Allies. Many of the essential characteristics of the modern monaural tape recorders can be traced back to these captured German machines. However, it all started with a man named Valdemar Poulsen, who received the first United States patent for a magnetic recorder.

Valdemar Poulsen was born in Copenhagen, Denmark, on November 23, 1869. His father was a judge in the highest court of Denmark. As a youngster, Poulsen was interested in physics and drawing and spent most of his time experimenting. In 1889, he enrolled in the University of Copenhagen to study for a degree in natural sciences. He did not stay at the university long enough to graduate but took a job in the technical department of the Copenhagen Telephone Company as a troubleshooter in 1893.

There, he had time to continue his experimenting, especially with the telephone that was still in its infancy. From these experiments came the idea of magnetic recording. He built a box containing two drums that wound and unwound a length of steel wire. As the wire moved from one drum to the other, it was pressed against the poles of an electromagnet. A microphone converted sound into electrical components that caused the electromagnet to magnetize the wire in localized areas. These recorded magnetized areas could be converted back to sound with the aid of a telephone. Poulsen called his device the Telegraphone. He filed a patent application in Denmark in 1898 and one year later in the United States.

In 1903, Poulsen and his associate, Peder O. Pedersen, granted permission to the American Telegraphone Company of Springfield, Massachusetts, to begin production of the Telegraphone. Poor amplification and poor management resulted in the company eventually being placed into receivership. The only other work Poulsen and Pedersen did on the Telegraphone was in 1907 when they improved the quality of recording by discovering d.c. biasing.

After 1900, Poulsen divided his attention between the Telegraphone and radio. In 1900, William Du Bois Duddell, a British electrical engineer, discovered a method of generating continuous electric waves by an electric arc. In 1903, Poulsen applied these continuous waves to a wireless transmitter to produce a wireless telephone. During World War I, several battleships carried these arc transmitters.

Poulsen received many honors. His Telegraphone won the Grand Prix at the Paris Exposition in 1900. In 1907, he received the Gold Medal from the Royal Danish Society for Science. In 1909, the University of Leipzig conferred upon him the honorary Doctor of Philosophy. He also received the Medal of Merit from the Danish government. Poulsen died on July 23, 1942, in Gentofte, Denmark.

Motion Picture

Most everyone knows who invented the phonograph and the electric light. However, few people realize that the same man also invented motion pictures. His name is Thomas Alva Edison and he received the first U.S. patent for a motion picture camera capable of taking a series of photographs that, when viewed in succession, produce motion pictures. The patent number was 493,426, issued March 14, 1893.

The history of motion pictures evolved from the thinking and work of many people. In 65 B.C., Lucretius, a Roman poet, discovered the principle of the persistence of vision. In 135 A.D., Ptolemy, a Greek astronomer, proved this same principle by experiments. In 1824, Peter Mark Roget, a British physician and scholar who compiled *Roget's Thesaurus*, presented a paper before the Royal Society in London entitled, "The Persistence of Vision With Regard to Moving Objects." The ideas presented are considered the scientific basis of the moving picture.

Toys played a major role in the early history of motion pictures. In 1826, Henry Filton, an English scientist, developed a toy called the thaumatrope that consisted of a paper disk with a drawing of a cage on one side and a drawing of a bird on the other. When the disk was whirled by strings, the bird appeared to be inside the cage. In 1832, Joseph Antoine Ferdinand Plateau, a Belgium scientist, developed the phenakistoscope, the first device for making drawings appear to move. In 1852, Franz von Uchatius, an Austrian field marshal, was the first to project drawings with a magic lantern, making them move on the wall by rapidly letting one succeed the other. In 1860, Pierre Hubert Desvignes of France developed the zeotrope. On February 5, 1861, a peepshow was patented by Samuel D. Goodale of Cincinnati, Ohio, as U.S. Patent No. 31,310. In 1861, Coleman Sellers, an American mechanical engineer, was the first to use photographs, rather than drawings, in a device called a kinematoscope. On April 23, 1867, U.S. Patent No. 64,117 was issued to William E. Lincoln of Providence, Rhode Island, for a machine to show animated pictures. He assigned it to Milton Bradley & Company, Springfield, Massachusetts, who made it as a toy. In 1868, John Barnes Linnett patented a device, called a kineograph that consisted of successive pictures of a single motion made in the form of a little book. When the edges were flipped rapidly,

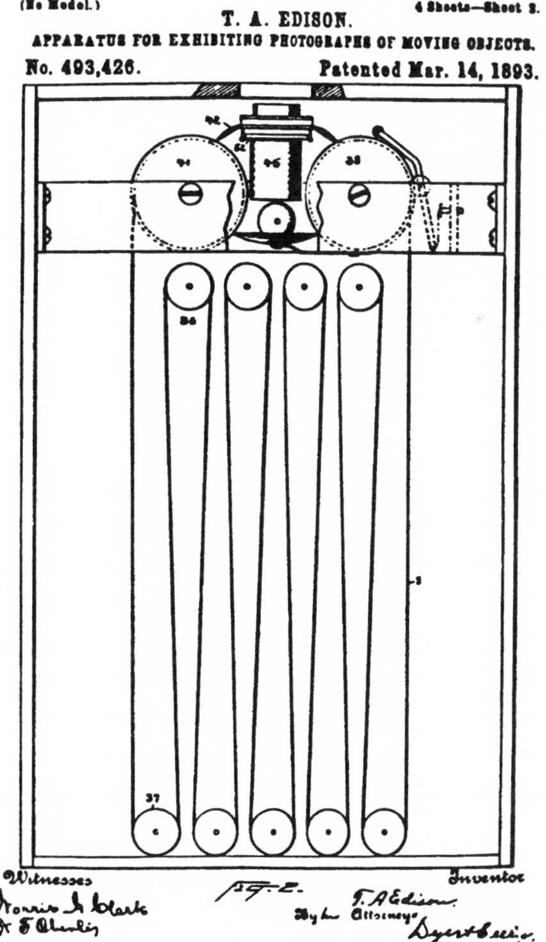

Thomas Alva Edison received the first U.S. patent for a motion picture camera in 1893

the pictures appeared to have motion. On August 10, 1869, O. B. Brown of Malden, Massachusetts, was granted U.S. Patent No. 93,594 for a moving picture projector that combined the principles of the phenakistoscope and the magic lantern. It was called a phasmatrope in which colored drawing transparencies were caused to revolve between an arc lamp and a lens.

The invention of photography was the catalyst that precipitated the advances in motion pictures. In 1870, Henry Heyl of Philadelphia, Pennsylvania, projected 18 photographic views of a waltzing couple on a screen as an illustration of reality. In view of the fact that photography with living models was used and that the pictures were viewed on a screen, this is probably the first demonstration of the motion picture as we know it today. In 1872, Eadweard Muybridge, an Englishman and photographer in the United States Coast and Geodetic Survey, was hired by ex-Governor Leland Stanford of California to prove by photography that a race horse, when running, has all four feet off the ground at one time. Muybridge set up 24 cameras on a race course in Palo Alto, California, attached threads to the shutters, and as the horse galloped past the row of cameras, it broke one thread after the other, operating the shutters. He continued for many years to experiment on animal locomotion, eventually publishing a book in 1878 called *The Horse in Motion.* He exhibited his photographs on a device called the zoopraxiscope, which had a capacity of 200 still photographs. In 1874, French astronomer Pierre Jules Janssen tracked the path of the planet Venus across the sun using a succession of photographs, which he exhibited later in movie form.

In 1885, the Reverend Hannibal Goodwin, a 65-year-old Episcopal Minister of Newark, New Jersey, discovered that celluloid was a proper base for photographic emulsions. Later, George Eastman of the Kodak Company, Rochester, New York, commercialized this discovery by cutting the celluloid into strips, placing a photo-sensitive emulsion on the strips, and then offering them for sale to the public in roll form. In 1886, Etienne J. Marey of France devised the first movie camera with film reels and shot a short movie of a bird in flight at 16 exposures per minute. In 1888, Augustin Le Prince, a French artist, built a camera and used film to shoot traffic moving across a bridge at Leeds. He and his brother were preparing to commercialize this idea when he mysteriously disappeared in 1890 on a train from Dijon to Paris. In 1890, William Friese-Greene and Mortimer Evans, both of England, made a camera capable of 10 exposures per second. However, it was Thomas Alva Edison who, on March 14, 1893, received the first U.S. patent for a movie camera for taking pictures, called the Kinetograph, and a device for viewing pictures, called the Kinetoscope.

The story of Thomas Alva Edison's life has been told many times but perhaps as interesting as the invention of the electric light and the phonograph is the story of his 691st U.S. patent, his invention of the movie camera and viewer. It all started when Edison heard a lecture in Orange, New Jersey, in February 1888, by Eadweard Muybridge on the subject of "Animal Locomotion." Muybridge later sent him over 100 photographs. These photographs probably sparked his interest in motion pictures inspiring him to put eyes to his phonograph for which sales now waned. In 1899, Edison attended the Paris Exposition where he saw motion pictures exhibited by Etienne Marey. This further peaked his interest in the subject.

Edison at first tried to put pictures on a cylinder like that of his phonograph but soon abandoned this idea. Then, in the summer of 1889, he read an advertisement by the Kodak Company that "Roll Photography," in which a layer of photosensitive emulsion was coated on a strip of flexible celluloid, was available to the public. He immediately ordered a 50-foot roll for $2.50 and he and his assistant, William Kennedy Laurie Dickson, started to build the first successful movie camera and viewer. In a short time, they had completed the Kinetograph for taking pictures and a peepshow device, called the Kinetoscope, for viewing the pictures. Edison then built the world's first motion picture studio in South Orange, New Jersey, called the Black Maria. The first public demonstration of the Kinetoscope by Edison was May 20, 1891, to the delegates of a convention of the National Federation of Women's Club in Orange, New Jersey. On March 14, 1893, he received U.S. Patent No. 493,426 for his invention. On April 14, 1894, he opened the Kinetoscope Parlor at 1115 Broadway in New York where a single viewer for a small price sat at a peephole to see the film. The performance lasted about 13 seconds and was regarded as a great success. By the mid 1890s, over 1,000 Kinetoscopes had been manufactured by Edison.

On June 6, 1894, a stenographer for the U.S. Treasury, C. Francis Jenkins, projected strips of film on a screen with a homemade projector in his father's shop in Richmond, Indiana. He repeated this viewing before a larger audience at the Cotton Exhibition in Atlanta, Georgia, in August 1895. When Edison heard about this, he bought the patent from Jenkins to eliminate any competition for his Kinetoscope. A short time later, Thomas Armat successfully projected pictures on a screen in Atlanta, Georgia, in September 1895. When the public heard about the Jenkins and Armat screen projected pictures, they began to demand this type of entertainment. So Edison bought the Armat invention, named it the Vitascope, and began to manufacture it under his own name. The first showing utilizing the Vitascope was April 23, 1896, at Koster & Bial's Music Hall in New York.

Edison's lawyers advised him to take out European patents on his Kinetoscope, but he refused saying that it wasn't worth the $150 cost. It was not long until some Europeans started to take advantage of this decision. In 1895, Robert W. Paul of England, Max and Emil Skladanowsky of Germany, and Louis and Auguste Lumiere of France built and improved Edison's Kinetoscope and began to show pictures on a screen before audiences in Europe. Soon, there was a demand throughout the world for moving pictures. On April 2, 1902, the first motion picture theater opened as the Electric Theater in Los Angeles, California. The first film to tell a story in pictures was made in the Edison studios in 1903. It was called *The Great Train Robbery* and was directed by Edwin S. Porter.

In 1908, Edison built up a trust, the Motion Picture Patents Company, to monopolize the industry. He bought all the motion picture patents he could. In 1917, the U.S. Supreme Court ordered the Motion Picture Patents Company dissolved because it violated the antitrust laws, but not before it made Edison a millionaire.

Edison received 1,093 U.S. patents for his various inventions. Certainly, one of these, No. 493,426 for the Kinetoscope, ranks high in importance. Other contributions to the film industry, besides being the first, was his use of celluloid as a film base, his 35 millimeter width, and his system of perforations to control the motion of the film.

Number One

United States Patent No. 1 was issued to John Ruggles, a senator from Maine, on July 13, 1836.

Between 1790 and July 4, 1836, the United States granted 9,802 patents. These patents originally did not contain an identification number and are usually referred to as "name and date" patents. Later, they were arbitrarily assigned numbers followed by X in the sequence in which they were issued. The Patent Act of 1836 established, among other things, a numbering system for issued patents for easier identification. This system started with number one on July 13, 1836, and remains in effect today after more than 5 million patents have been issued. The only deviation was in 1861 when for an unknown reason a new numbering system was adopted and then abandoned in the same year. Thus, some United States patents issued in 1861 may have two numbers, the higher number being the correct one.

John Ruggles was born in Westboro, Massachusetts, on October 8, 1789. After attending the common schools in Westboro, he enrolled in Brown University in Providence, Rhode Island. He graduated from Brown in 1813 and then studied law. After being admitted to the bar, he started practicing law in Skowhegan, Maine, in 1815. In 1817, he moved his law practice to Thomaston, Maine.

Ruggles was elected as a member of the State of Maine House of Representatives in 1823, remaining there until 1831. He served as speaker from 1825 to 1829 and again in 1831. He was appointed as justice of the Supreme Court of Maine in 1831. He was elected as a Democrat to the United States Senate and served from January 20, 1835 to March 3, 1841. He ran again in 1840, but was defeated. He resumed his law practice in 1841, in Thomaston, Maine. He was also active as an inventor, orator, and writer. He died on June 20, 1874, in Thomaston and was buried in Elm Grove Cemetery.

Being an inventor, an attorney, and a member of the United States Senate, Ruggles turned his attention toward the patent laws. He soon saw the need for reform of these laws. Since 1793, the State Department had issued patents for 40 years without any examination of their merits. On December 31, 1835, he proposed to the Senate that a committee be appointed to study and make recommendations for any revisions they deemed necessary in the patent laws. The Senate did appoint such a committee with Ruggles as the chairman. On April 28, 1836, the committee, after carrying out their obligation by studying the present patent laws, submitted their report and incorporated their findings into Senate Bill 239.

Congress passed the Patent Act of 1836 on July

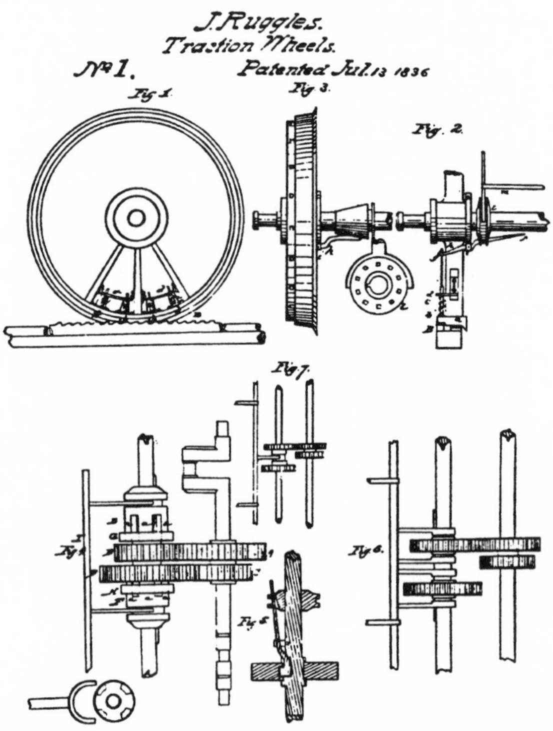

United States Patent No. 1, for traction wheels, issued to Maine senator John Ruggles on July 13, 1836.

4, 1836, which eliminated the registration system and restored the examination of patent applications. It also created a Patent Office, established a Scientific Library, initiated the office of Commissioner of Patents, and put into effect a numbering system for issued patents.

Shortly after the Patent Act of 1836 went into effect, United States Patent No. 1 was issued on July 13, 1836, to John Ruggles for improvements in locomotive traction.

Nylon

The first U.S. patent for a truly man-made synthetic fiber, Nylon 66, was awarded to Wallace Hume Carothers on February 16, 1937. The number was 2,071,250.

The chief textile fibers until the 1700s were flax, wool, and silk. After the English textile manufacturers developed machines to spin and weave cotton and Eli Whitney invented the cotton gin in 1793, these materials were replaced by cotton as the main textile fiber. In 1884, another fiber found its way to the marketplace when Frenchman Hilaire de Chardonnet patented the first practical synthetic fiber, called artificial silk, later named rayon in 1924. But it was Wallace Hume Carothers's invention of nylon on February 16, 1937, and the tremendous work of the du Pont Company that revolutionized the textile fiber industry in the 1940s.

Wallace Hume Carothers was born in Burlington, Iowa, on April 27, 1896. His father, Ira Hume Carothers, was a teacher and later vice president of Capital City Commercial College in Des Moines, Iowa. Young Carothers received his elementary and high school education in Des Moines. He enrolled in the Capital City Commercial College in 1914, and after a year transferred to Tarkio College in Tarkio, Missouri, to pursue scientific studies. He earned college expenses by means of teaching assistantships in accounting, English, and chemistry. He received the BS degree from Tarkio College in 1920, and the MS degree in organic chemistry from the University of Illinois in 1921. He then served for a year as a chemistry instructor at the University of South Dakota.

In 1922, Carothers returned to the University of Illinois as a doctoral candidate. His doctoral thesis dealt with the catalytic hydrogenation of aldehydes over a platinum catalyst, and was presented in 1924 under the direction of Professor Roger Adams. During this time, he published two papers on theoretical organic chemistry in the *Journal of the American Chemical Society*. He remained at the University of Illinois for two years as instructor in organic chemistry and became interested in the study of polymers of high molecular weight. He went to Harvard University as an instructor in organic chemistry in 1926.

In 1927, Dr. Charles M. A. Stine, head of du Pont's chemical department, persuaded the executive committee to set aside $250,000 to fund a program of fundamental research. Dr. Wallace H. Carothers was offered the direction of this research group. He accepted and joined the firm in February 1928. His early work at du Pont centered around the reaction of a bifunctional organic acid and a bifunctional alcohol to form a long-chain polyester in

Feb. 16, 1937. W. H. CAROTHERS 2,071,250
LINEAR CONDENSATION POLYMERS
Filed July 3, 1931

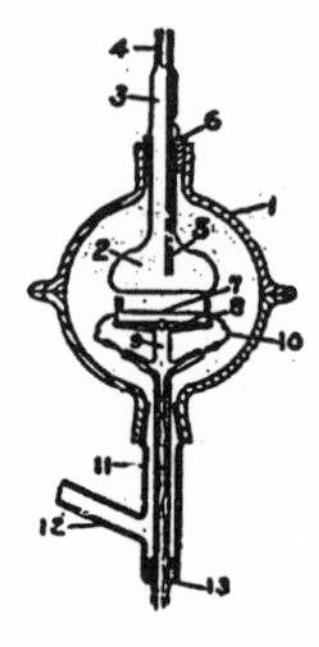

Wallace H. Carothers, Inventor

By A. F. Miller
Attorney

Linear Condensation Polymers, patented by W. H. Carothers, issued February 16, 1937

which the acid and the alcohol are joined end to end. He also investigated certain derivatives of vinylacetylene from which a rubber-like polymer was produced. This research later led to the commercial development of a synthetic rubber known as neoprene.

In 1930, Dr. Julian W. Hill, working under the direction of Carothers, noted that the polymer formed from the polymerization reaction of ethylene glycol and sebacic acid could be drawn out in the form of a long fiber and that this fiber when cold could be further drawn out to several times its original length, resulting in an increase of the fiber's strength and elasticity. However, this particular fiber was easily softened by hot water, and, therefore, was not practical. Many other new polyesters were made, but none had the characteristics necessary for a commercially acceptable fiber. Carothers gave up the polyester project for several months and directed his attention to other research.

Dr. Elmer K. Bolton, who had succeeded Dr. Stine as director of the du Pont chemical department, urged Carothers to continue his work on superpolymers from which a commercially useful fiber might be made. Carothers then turned his attention to polyamides, made in the same way as the polyesters except that a bifunctional amine is used instead of a bifunctional alcohol. After preparing several hundred different polyamides, Carothers and his associates on February 28, 1935, found that the polymer made by the reaction of hexamethylenediamine and adipic acid proved to be the most promising for having the properties necessary for forming a superior fiber. This product was called Nylon 66.

To make the fiber commercial, du Pont spent much time and money overcoming many problems. First, the two raw materials, hexamethylenediamine and adipic acid, were not available in sufficient quantity. The chemists and chemical engineers at du Pont went to work on this problem and successfully produced these two materials in large amounts. Five years and $27 million after the invention of Nylon 66, it was announced that a large plant was to be built at Seaford, Delaware. It started operation on December 15, 1939, with a production capacity of 8 million pounds of nylon per year. Nylon stockings went on sale commercially for the first time on May 15, 1940, and in the first year some 64 million pairs were sold. During World War II, nylon's use was limited to the military, principally for parachutes.

In 1936, Carothers was elected to membership in the National Academy of Sciences, the first research organic chemist of a manufacturing company to be so honored.

Carothers did not live to see the commercial results of his invention. Subject to periods of deep depression, he ended his life on April 29, 1937, in Philadelphia. His ashes were buried in Glendale Cemetery in Des Moines, Iowa.

Phonograph

Millions of people have enjoyed the phonograph, and perhaps many know that it was invented by Thomas Alva Edison. He received United States Patent No. 200,521 for a phonograph on February 19, 1878.

Little work had been done on the phonograph before Edison. In 1807, Thomas Young of England explained how to record vibrations of a tuning-fork on the smoked surface of a rotating drum. In 1857, Leon Scott, also of England, invented what he called the phonautograph, an instrument that had the capacity to register vibrations made by the human voice or musical instruments. This device was similar in appearance to an ear trumpet or an open-ended cone. It had a membrane that was stretched across the smaller opening and attached to the membrane was a stiff hair stylus. Any sound entering the larger opening would vibrate the membrane which in turn caused the stylus to produce a line corresponding to the vibration on a revolving cylinder covered with blackened paper. In 1863, Fenby received British Patent No. 101 of 1863 for a device for recording the vibrations caused by the piano. In 1877, Charles Cros received French Patent No. 124,213 for a device that also recorded sound vibrations by means of a metal stylus attached to the center of a diaphragm that vibrated with sound to trace a wavy line on a lamp-blackened disk or cylinder. These devices could record only sound; none could reproduce the recorded sound until Edison's invention of the phonograph in 1877.

Thomas Alva Edison was a prolific inventor who produced a diversity of inventions. He obtained 1,093 United States patents. For 65 consecutive years, from 1869 to 1933, he obtained one or more patents in every year, the last four being awarded posthumously. In 1882 alone, he was granted 75 patents. Of his many inventions, the phonograph is regarded as one of his greatest achievements.

Edison did not set out to invent the phonograph. Rather, it came about as a result of his work in telegraphy. In 1877, Edison was working on a device that would permanently record the dot-dash pattern of Morse code by making indentations on paper wrapped around a rotating cylinder. His purpose was to play back this dot-dash pattern at a speed faster than the recording speed so that messages might be transmitted more rapidly than those transmitted by hand, thus permitting more messages to be sent in a given time. He observed that when the cylinder with

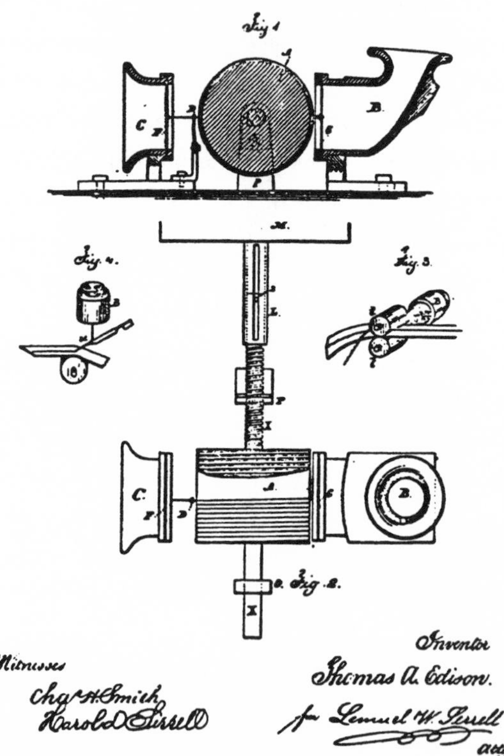

Regarded as one of Thomas Alva Edison's greatest achievements, the patent for the phonograph, or speaking machine, was granted February 19, 1878

the indented paper was rotated at a high speed, a noise that resembled speech came from the machine.

During this same time, Edison was also experimenting with a telephone diaphragm that had a sharp stylus attached to the rear side. As he spoke against the front of the diaphragm, the stylus was driven into his finger. On August 12, 1877, combining the observation of the sound that came from the rotating indented paper with the idea that the telephone diaphragm vibrated with sufficient energy to make an indented record, he drew a crude pencil sketch of a device and handed it to one of his assistants, John Kruesi. Instructions at the bottom read: "Kruesi: Make this—Edison." Kruesi was also instructed not to spend more than $18 in making it.

Kruesi, a Swiss watchmaker, immediately went to work following the idea of Edison's rough sketch. In a short time, he had built a device consisting of a long brass cylinder covered with tinfoil and mounted on a feed screw. It was to be turned by a hand crank. On each side of the cylinder was a fixed diaphragm having a needle that could be adjusted against the tinfoil and actuated by the human voice to make indentations in the tinfoil as the cylinder revolved and moved past the needle. The result was a helical groove with indentations in the tinfoil. For playback, another fixed diaphragm having a rounded needle was allowed to follow the helical groove as the cylinder rotated.

On December 6, 1877, Kruesi delivered the requested device to Edison. The entire laboratory staff gathered around the table as Edison turned the cylinder and spoke into the mouthpiece: "Mary had a little lamb." They anxiously watched as Edison adjusted the reproducing needle and again rotated the cylinder. From the machine came the words: "Mary had a little lamb." The staff was amazed as was Edison himself. They spent the night, taking turns speaking and listening to their own voices. Edison named his talking machine the phonograph.

The next day Edison demonstrated his phonograph in New York before Alfred E. Beach, editor of the *Scientific American*. Immediately, the news spread throughout the world about this astonishing invention. On December 24, 1877, Edison filed an application for a patent. The patent examiner in charge of the application could find no prior art, and the patent was issued on February 19, 1878. Edison was invited to the White House to demonstrate his talking machine before President and Mrs. Rutherford B. Hayes. Thousands of people came to the Menlo Park Laboratories to see and listen to the phonograph. So great was the demand that Edison decided to manufacture and sell his tinfoil phonographs. He established the Edison Speaking Phonograph Company and secured the services of Sigmund Bergmann & Company of New York to manufacture the machine. Although it was popular at first, sales began to drop because of the difficulty of the hand operation. The public began to lose interest, so Edison turned his attention to the incandescent lamp and did not work on the phonograph again for nine years.

In the meantime, others took up the work and began to improve the phonograph. On May 4, 1886, Chichester Bell and Sumner Taintor, both of Washington, D.C., were granted United States Patent No. 341,214. Their device removed at varying depths part of the recording medium instead of indenting it as did Edison's. On November 8, 1887, Emile Berliner, also of Washington, D.C., was issued United States Patent No. 372,786 for the gramophone in which the cylinder was replaced with a flat disk. In 1887, Edison resumed work on the improvement of his phonograph. He replaced the tin foil with a cylinder of wax as the recording medium. He also replaced the hand crank with a clock-spring mechanism for rotating the cylinder. From this humble beginning of the Bell, Taintor, Berliner, and Edison patents grew two recording giants, Columbia and RCA Victor. Although the Berliner disk eventually replaced the Edison cylinder, it remains a fact that Thomas A. Edison was the first to record and reproduce sound and was the first to receive a United States patent for the phonograph.

Phonograph Records

The multimillion dollar phonographic record industry got its start when Emile Berliner received U.S. Patent No. 372,786 for the Gramophone on November 8, 1887.

In 1857, Leon Scott, a French inventor constructed a device called the phonautograph, which traced lines representing sound on the smoked surface of a rotating cylinder. He never perfected his invention. In 1877, Thomas A. Edison of the United States patented the first workable phonograph, which used a tinfoil-coated cylinder to record and reproduce the sound. In 1878, Charles Cros of France published a paper describing a means of recording and reproducing sound but never actually built such a machine. In 1886, Chichester Bell and Charles Sumner Tainter improved the Edison machine by substituting a wax cylinder. They called the apparatus a Graphophone. In 1887, it was Emile Berliner who revolutionized the phonographic industry by recording on a flat metal disk from which copies could be made.

Emile Berliner was born May 20, 1851, in Hanover, Germany. His father, Samuel M. Berliner, was a merchant and a Talmudic scholar. Emile graduated in 1865 from the Samsonschule in Wolfenbuttel. He was one of 11 children, and at the urging of his parents, found work as a printer's devil and then as a clerk in a dry-goods store in Hanover. When he was 16, he devised the first of his many inventions, a weaving machine.

In 1870, a family friend, Nathan Gotthelf, persuaded Berliner to come to the United States. He came to Washington, D.C., and worked in the dry-goods store owned by Gotthelf. In 1873, Berliner moved to New York where he worked selling glue, painting backgrounds into tintype portraits, and giving German lessons. A short time later, he moved to Milwaukee, Wisconsin, and for the next year worked as a clothing salesman. He then returned to New York and worked in the laboratory of Constantine Fahlberg, who later discovered saccharin. In 1876, he took out citizenship papers and returned to Washington and his old job.

In 1876, Berliner traveled to Philadelphia where Alexander Graham Bell was exhibiting his telephone at the International Centennial Exhibition. After seeing the telephone in operation, Berliner became interested in telephony. He equipped his room in his Sixth Street lodgings with a maze of

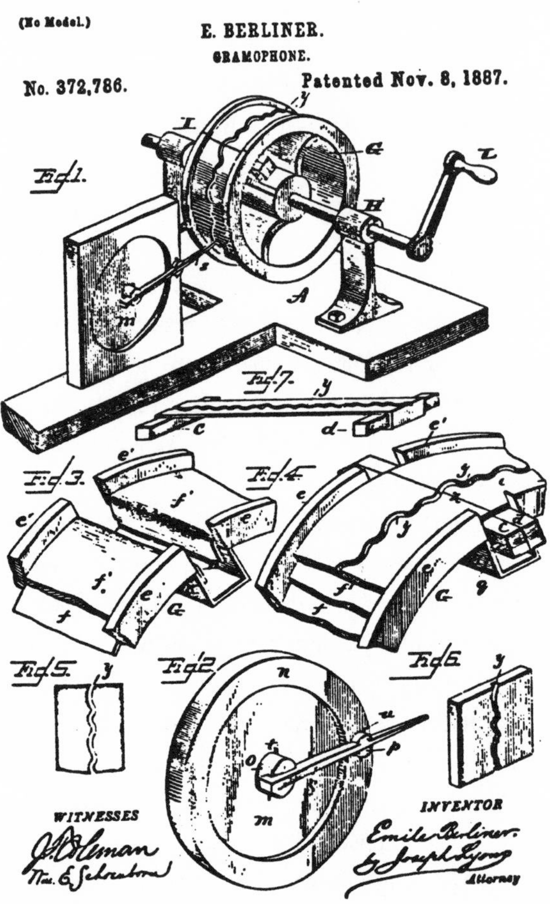

The beginning of the record industry—Emile Berliner's Gramophone patent, granted November 8, 1887

wires, battery jars, transmitters, and receivers. At the suggestion of his friend, Alvan S. Richards, chief operator of the Washington fire alarm telegraph office, he invented a telephone transmitter, or microphone, that worked on a principle different from that of Bell's. Berliner did not have enough money to pay a patent attorney, so he filed a caveat himself on April 14, 1877. In March of 1878, the Patent Office declared an interference on all microphone applications, including those of Bell, Gray, Dolbear, Richmond, Holcombe, Edison, and Berliner. After 14 years of litigation, United States Patent No. 463,569 was awarded to Berliner on November 17, 1891.

In 1878, Berliner sold his microphone invention to the American Bell Telephone Company and moved to Boston to become the company's chief instrument inspector. While there, he improved the microphone by incorporating an induction coil transformer. In 1881, he made a trip back to Germany, where he set up his brothers in business manufacturing his microphone for European markets. In 1883, he returned to Washington to live the rest of his life.

In 1887, Berliner turned his attention to the phonograph. He turned a front room of his house into a laboratory and by November of that year had developed and patented a new device to record and reproduce sound that he called the Gramophone. His system differed from that of Edison's in two ways. He registered the sound waves by having the needle move laterally instead of up and down as did the Edison needle. He also substituted a flat disk for the cylinder of Edison. This system enabled Berliner to make as many copies as he liked from a single original recording. These copies became known by 1896 as "records." He first demonstrated his machine before the Franklin Institute in Philadelphia on May 16, 1888. The first machines were produced by the United States Gramophone Company of Washington, D.C., in 1894. The first shellac records were made by the Berliner Gramophone Company of Philadelphia in 1897, from a composition prepared by the Durinoid Company of Newark, New Jersey.

Berliner's original machine was first operated with a handcrank that was later replaced by a small motor. He took his machine to Eldridge Johnson, owner of a small machine shop in Camden, New Jersey, to have the motor improved. In 1897, Berliner and Johnson joined forces and founded the Victor Talking Machine Company. From this company came the famous trademark, His Master's Voice, an adaptation from Francis Barraud's painting of a fox terrier listening to the horn of a Gramophone. From this company also came the Victrola to play the new records.

In 1913, Berliner received the Elliott Cresson Medal from the Franklin Institute. In 1919, he and his son Henry A. Berliner designed, built, and flew a helicopter. On August 3, 1929, Emile Berliner died at his home in Washington, D.C., and is buried in Rock Creek Cemetery.

Plant Patent No. 1

The recipient of United States Plant Patent No. 1 was Henry F. Bosenberg on August 18, 1931.

The various patent acts before 1930 did not include plants as patentable entities. In 1889, the United States Patent Office rendered a decision that denied a patent on a fiber found in a pine-tree needle, stating that it was a product of nature and as such was not patentable. This decision formed a legal precedent for years in denying patents on plants. The prospect for securing patents for plants remained dim until the late 1920s and early 1930s, when three men fought for the passage of a law enabling plants to be patented.

One member of this trio was Luther Burbank, the world famous horticulturist, who had a working agreement with Stark Brothers Nurseries of Louisiana, Missouri, from 1893 to the time of his death in 1926. During this time, he introduced more than 800 new varieties of trees, vegetables, fruits, and flowers but had never received a United States patent for his work because of the absence of an enabling law. Another member was Paul Stark, from Stark Brothers Nurseries, who had become a lobbyist for the American Association of Nurserymen to secure passage of a plant patent act. The third member was Thomas Edison, a friend of both Burbank and Stark, who thought it was inequitable to be able to patent a new gadget but not a new fruit. Stark used the testimony of both Burbank and Edison in lobbying Congress for passage of some kind of law protecting plant breeders.

The words of these three men apparently had an effect on Congressional ears. On February 11, 1930, Senator John G. Townsend of Delaware and Representative Fred S. Purnell of Indiana introduced identical bills in the Senate and the House for plant patents. After little debate, an amended bill passed in the Senate on May 12, 1930, and the following day it was passed in the House. The Townsend-Purnell Plant Patent Act of 1930 became law after President Hoover's signature on May 23, 1930. The new law covered only asexually propagated plants and excluded seed-propagated plants. Asexually propagated plants are those that are reproduced by the rooting of cuttings and by layering, budding, and grafting.

The first to file an application for a plant patent

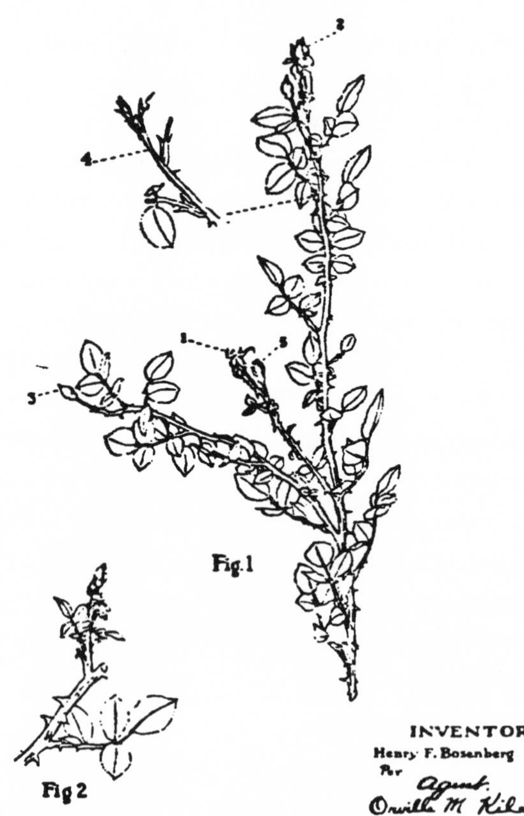

Henry F. Bosenberg's patent for the climbing or trailing rose was the first plant patent granted in the United States

was Frank Spanbauer of North Kansas City, Missouri, on July 1, 1930, but he was not the first to receive a patent. The first application to receive a plant patent, on August 18, 1931, was filed on August 6, 1930, by Henry F. Bosenberg of New Brunswick, New Jersey. The Spanbauer application matured into Plant Patent No. 2 on October 13, 1931, but has the distinction of being the first patent drawing to be issued and reproduced in color. It is interesting that although Luther Burbank received no patent while he was alive, his widow Elizabeth Burbank, as executor of his estate, received in his name Plant Patent No. 12 for a plum tree on April 5, 1932.

Plant patent applications and drawings are filed in duplicate in the United States Patent and Trademark Office. The duplicates are forwarded to the Agriculture Research Service, Horticultural Crops Research Branch, Department of Agriculture, for a report on the novelty of the plant. Once this report is received, the Patent and Trademark Office will then take action on the application.

This is what happened to the application of Bosenberg. He was not a plant breeder, as was Burbank, but a landscape gardener in New Brunswick, New Jersey. He ordered several roses for use in his landscape work. Some of these roses were Van Fleet roses, originated by Dr. Walter Van Fleet, a well-known plant breeder. Bosenberg experimented with the Van Fleet rose for a couple years and observed that they continued to bloom after the original blooms had passed. He propagated these experimental roses by budding or grafting and observed that the young plants had everblooming characteristics. He continued propagation studies for four years and observed no tendency of the plants to revert to the old rose that bloomed only once a year. On August 6, 1930, he filed a patent application, calling his rose New Dawn. The application became Plant Patent No. 1 on August 18, 1931, after Department of Agriculture reports and an affidavit by the applicant stressing the everblooming characteristics.

In 1968, the seed industry attempted to expand the 1930 Plant Patent Act to secure patent rights for seed-propagated plants. This attempt was opposed by the Johnson Administration, the Senate Judiciary Committee, the Farm Bureau, and the Crop Science Society of America. Facing such opposition, the seed industry changed its tactics and began to pursue a patent-like system in which a 17-year patent seed certificate would be issued by the Department of Agriculture rather than by the Patent Office. This bill, known as the Plant Variety Protection Act, was introduced into Congress in late 1969 by Representative Graham Purcell of Texas and Senator Jack Miller of Iowa. After little opposition, mainly from the Campbell Soup Company, the bill was amended to exclude tomatoes, peppers, okra, celery, cucumbers, and carrots. On December 9, 1970, Congress passed the Plant Variety Protection Act, and on December 24, 1970, President Nixon signed the bill into law.

The seed industry then started lobbying Congress to include the six vegetables that had been excluded in the 1970 Plant Variety Protection Act. In January 1979, Representative Kika de la Garza of Texas and Senator Frank Church of Idaho introduced bills into Congress to amend the 1970 Act to include the six vegetables and to change the 17-year patent seed certificate to an 18-year certificate. After much controversy and many delays, the amendment was finally passed and sent to the White House on December 8, 1980. President Carter allowed the bill to become law in late December 1980.

The Plant Variety Protection Office of the United States Department of Agriculture has issued hundreds of patent seed certificates, the major holder being the Asgrow Seed Company, owned by the Upjohn Company, Kalamazoo, Michigan. The United States Patent and Trademark Office has issued hundreds of plant patents, 70 percent of which have been for roses and other flowering plants and shrubs. It all started with the issuance of Plant Patent No. 1 to Henry F. Bosenberg, a landscape gardener of New Brunswick, New Jersey.

Plow

One of the most needed agriculture implements in the early history of American farming was the plow. The early settlers recognized this need and responded by inventing and securing hundreds of patents. The first United States patent for the plow was issued to Charles Newbold on June 26, 1797.

The history of the plow, like that of the wheel, is veiled in antiquity. It probably started with ancient Egypt, whose people used plows made of wood that were little more than a crooked stick cut from the fork of a tree and sharpened at one end. The Greeks and the Romans used the ard, which was a metal-tipped but lightweight wooden implement pulled by oxen, sometimes called a scratch plow. The Romans added a wooden moldboard to turn a furrow and later added handles and landslides for guiding the plow. In 1700, in Holland, the Rotherham plow, which was an improvement over the Egyptian and Roman plows in that it was smaller and lighter and had a more efficient moldboard, was built. In 1720, the first English patent for a wooden moldboard covered with iron was issued to Joseph Foljambe.

In 1767, James Small introduced the Rotherham plow to Scotland and in 1785, added cast iron moldboards and shares to this plow. The first cast iron plow used in America was imported from Scotland after the Revolutionary War and was the invention of James Small. In 1785, Robert Ransome of England patented a cast iron share whose surface was hardened by chilling. This was an improvement over the Small share in that it remained sharp after long use, whereas the share of Small's plow became dull after short use.

From 1788 to 1793, Thomas Jefferson studied, experimented, and wrote about the proper shape of the moldboard. He never filed for a patent on any of his inventions, but his moldboard designs found their way into many later developed plows.

In 1797, Charles Newbold received the first United States patent for a cast iron plow. It was cast in one piece except for the handles and beam. Many improvements on the Newbold plow were made in

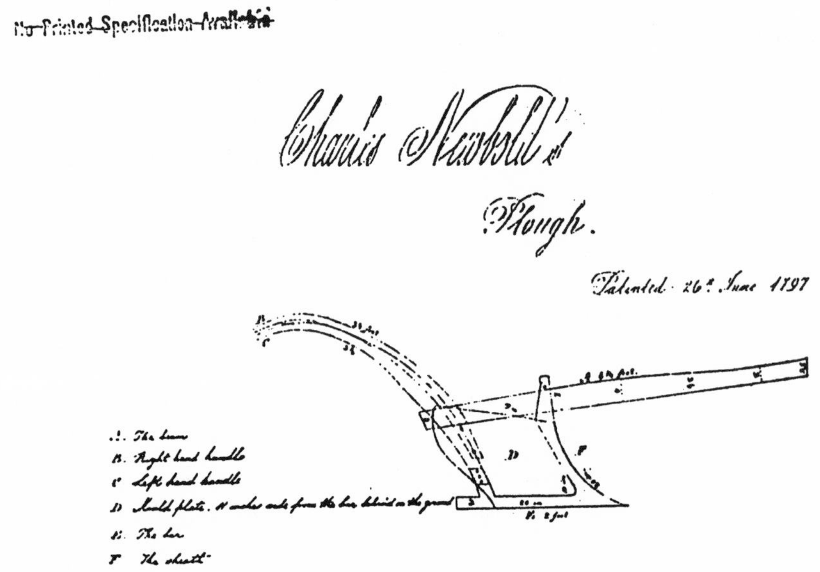

Though many patents for plows were granted throughout American history, the first was secured by Charles Newbold on June 26, 1797

the next few years. In 1807, David Peacock of New Jersey patented a plow that resembled Newbold's except that the point, share, and moldboard were cast separately and later fastened together. Newbold sued Peacock for infringement and collected $1,500. In 1819, Jethro Wood of Scipio, New York, introduced interchangeable parts. If one part wore out or became damaged, it could be replaced without buying the whole plow. He also invented the coulter which cuts the soil before the share strikes it. Wood's plow became very popular, and he became the victim of infringers. He spent all his royalties in defending his patent.

In 1833, John Lane of Chicago and John Deere of Moline, Illinois, unknown to each other, began to make steel-plated plows on which the share and moldboard were covered with sheets of steel made from discarded circular and hand saws. Deere, an astute businessman, formed John Deere & Company, and began to manufacture steel plows that became internationally known. He was the first inventor to make money on the plow.

Charles Newbold was born in the township of Chesterfield, Burlington County, New Jersey, in 1780. Little is known of his early life except that he developed an early interest in farming, and farming involved plowing. The plow had changed very little over the centuries until Newbold set his mind to improve this basic tool of the farmer. His aim was to build a plow that could be handled by one man and a team of oxen.

Between 1790 and 1796, Newbold experimented on a cast iron plow. When he decided to test his experiments on a larger scale, he had Benjamin Jones at the Hanover furnace in Burlington County cast the share, landslide, and moldboard in one piece. He attached handles and a beam and received permission to try it out in an orchard belonging to General John Black. The trial attracted a large audience of New Jersey farmers to watch the demonstration. The plow performed as Newbold had hoped it would, but its neat furrows failed to convince the watching farmers that a cast iron plow would not poison the ground and cause weeds to grow.

However, such superstitions did not dissuade Newbold from applying for a patent in 1797. A patent was granted on June 26, 1797, and signed by President John Adams, Secretary of State Timothy Pickering, and Attorney General Charles Lee. Even armed with this patent and the proof that his fields where his plow had turned the sod were not poisoned, Newbold could not convince the farmers of his area that his plow was in their best interest. After spending more than $30,000 of his own money in improving and introducing his plow to no avail, he finally gave up in disgust. He died at Cornwall, New York, without his plow becoming more than a museum piece.

Presidential Patent

The first and only United States patent issued to an inventor who later became president of the United States was No. 6,469 on May 22, 1849. The inventor was Abraham Lincoln.

The United States and Trademark Office has issued more than 5 million patents. Many of the inventors can be instantly recognized, such as Orville and Wilbur Wright, Thomas Edison, George Eastman, Samuel Morse, and Alexander Graham Bell. However, others may come as a complete surprise, such as Danny Kaye, Hedy Lamarr, Lillian Russell, Zeppo Marx, Edgar Bergen, Lawrence Welk, Lee Trevino, Mark Twain, and Harry Houdini.

Perhaps the most unique patent ever issued by the United States Patent Office was United States Patent No. 6,469 on May 22, 1849. It was issued to the only inventor who later became president of the United States. At least three men who served as president were inventors. George Washington invented a seeding plow and a wine coaster. Thomas Jefferson invented a swivel chair, a folding buggy top, a three-legged stool, a plowshare, and a writing desk with a folding top. However, neither of these men applied for a patent to their inventions. The only inventor who served as president to receive a patent for his invention was Abraham Lincoln.

Lincoln probably conceived his invention as a young man while on a trip by flatboat down the Ohio and Mississippi Rivers to New Orleans, and observed the problems steamboat navigators had on the shoals. His idea was to provide several buoyant chambers to the sides of the vessel that could be lowered into the water. These chambers once filled with air would allow the vessel to be buoyed and to pass over the shoal. Lincoln whittled a model with his own hands and secured the services of Z. C. Robbins, a patent attorney in Washington, D.C., to file and prosecute his patent application.

Although this invention alone had little, if any, bearing on Lincoln's being elected president, the fact

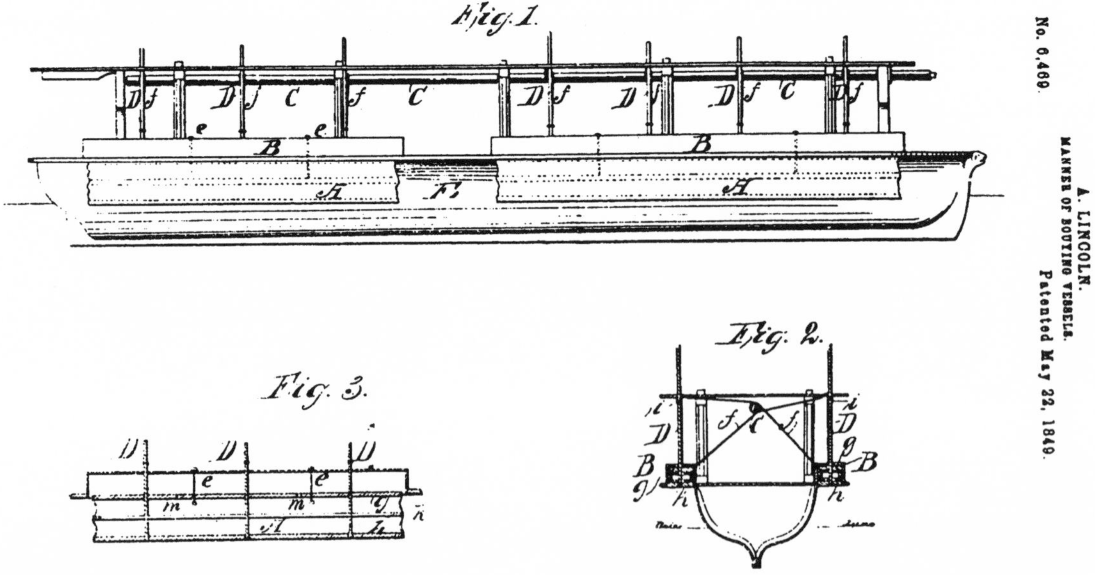

This patent demonstrating a manner for buoying vessels was granted to inventor Abraham Lincoln on May 22, 1849

that he had an inventive mind and an interest in mechanical things served him well while he was in office.

It was President Lincoln who pressured the Union admirals to accept the plans of John Ericsson to build an iron-clad warship having a revolving armored gun-turret. This vessel, called the *Monitor,* damaged the Confederate warship *Merrimac* to such an extent that the surviving wooden ships of the Union were saved from destruction.

Lincoln was also president when Christopher Spencer explained and demonstrated his invention of the first repeating rifle. A target was erected where the Washington Monument now stands, and after a few shots were fired, Lincoln determined that the gun was extremely practical. During the Civil War, through Lincoln's influence, more than 20,000 of these rifles were issued to the Union Army.

Although Lincoln's invention never enjoyed any commercial success, his ability to appreciate and evaluate new inventions did much to determine the future of this country.

Printing Press

The first United States patent, No. 5,199, for a rotary printing press was issued to Richard March Hoe on July 24, 1847.

Long before 1847, the Chinese had invented paper and block printing in about 105 A.D. This method entailed writing characters with a brush on a sheet of paper. While the characters were still wet, the paper was laid face down on a smooth board to transfer the ink lines. Then everything except the ink lines was cut away. The next improvement in printing came in the 1450s when Johannes Gutenberg of Germany invented the movable type press using separate pieces of raised metal type for each letter. The press was operated by manually turning a large wooden screw. In the 1450s, 300 copies of the famous 1,282-page *Gutenberg Bible* were printed.

The first print shop in North America was established in Mexico City in 1539, and was set up by Juan Pablos, an Italian. The first printing press in the American colonies was established in Cambridge, Massachusetts, in 1639 by Stephen Daye and his son Matthew. Perhaps the most famous print shop in early America was the one in Philadelphia run by Benjamin Franklin.

In 1790, William Nicholson was awarded British Patent No. 1,748, based on a rotary principle by which type and paper, each on separate cylinders, revolved together. In 1810, Friedrich Konig, a clockmaker from Saxony, Germany, invented a steam-powered cylinder press used in combination with a flat bed that was able to print 1,000 impressions per hour. This invention received British Patent No. 3,321 and was first used to print the London *Times.* It is considered to be the forerunner of the modern printing press.

In 1817, George Clymer of Philadelphia invented a printing press that was operated by hand. It was called the Columbian Press and he received British Patent No. 4,174 for this invention. This press

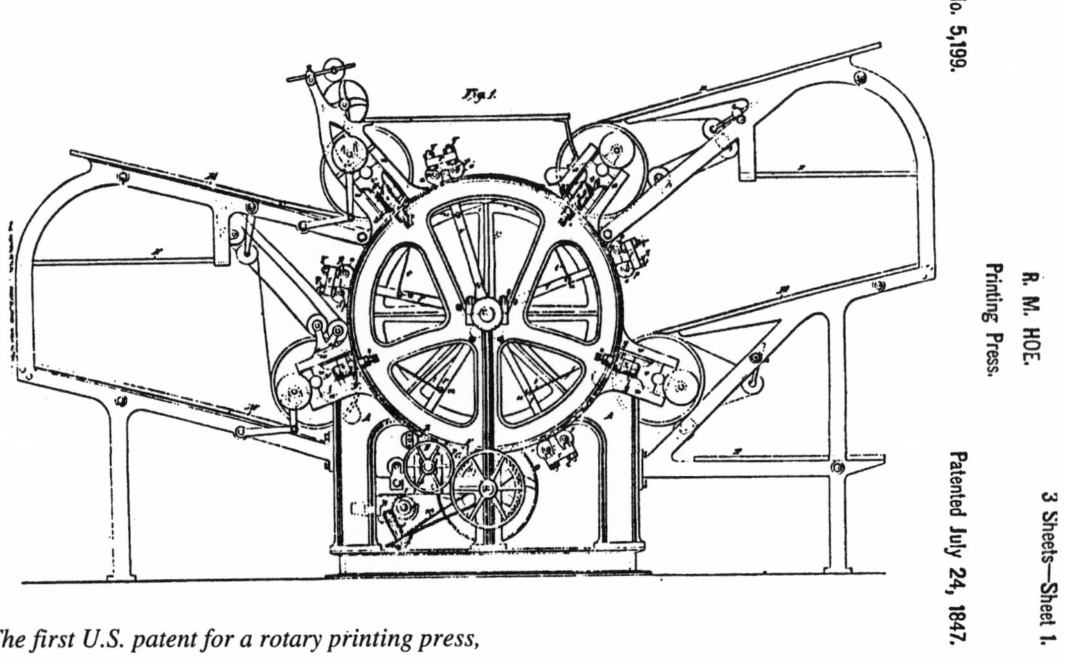

The first U.S. patent for a rotary printing press, issued to Richard March Hoe on July 24, 1847

eliminated the wood screw and substituted a set of integrated levers and toggle-jointed bars. In 1829, Samuel Rust patented in the United States the Washington Press. In this press, the platen is forced down by a lever and is raised by springs on each side. However, none of these presses was fast enough for the rapidly expanding newspaper business. It remained for Richard Hoe and his rotary press to meet this demand.

Richard March Hoe was born in New York City on September 12, 1812. His father Robert Hoe, who came to the United States from England in 1803, was a printer and inventor and owned a printing business in New York. The first years of Richard's life were spent in school. At the age of 15, he entered his father's printing establishment on Gold Street. Robert Hoe retired from the business because of failing health in 1830, and the management was turned over to Richard and his cousin Matthew Smith.

Shortly after assuming responsibility of the business, Hoe enlarged the operation on Gold Street by buying the Washington Press from Samuel Rust, and adding the manufacture of steel circular saws. The Hoe Company had for years made and sold the cylinder press, and in 1837, Hoe improved this machine by introducing the double small-cylinder press. In 1844, Hoe invented and patented the single large-cylinder press, the first flatbed and cylinder press ever used in the United States. For this invention, he received United States Patent No. 3,551. The company made and sold hundreds of these presses.

The demand by the newspaper industry for faster printing presses led Hoe to invent the Hoe type-revolving machine. It was patented on July 24, 1847, and received United States Patent No. 5,199. This machine had a central horizontal cylinder about 4 1/2 feet in diameter containing securely fastened type. Around this central cylinder, and parallel with it, were grouped from four to 10 impression cylinders. As the sheets of paper were fed, automatic grippers carried them between the impression and central cylinders, where they were printed. The machine was capable of printing 20,000 complete newspapers hourly. The first of these presses was placed in the office of the *Philadelphia Ledger* in 1847. By 1860, newspapers in all the larger cities had installed this machine.

Under Hoe's leadership, the company prospered. In 1859, the company purchased the Isaac Adams Press in Boston and shortly thereafter built a new factory covering an entire block at the corner of Broome and Sheriff Streets in New York. In 1871, Hoe and one of his partners, Stephen D. Tucker, designed and built a web press that printed from a continuous web, or roll, of paper. This press was first introduced by William Bullock in 1865. The Hoe web press was first placed in the office of the *New York Tribune* and produced 18,000 papers an hour.

Hoe was known as "Colonel," a title that came from his service in the National Guard. He built a large house in Westchester County, New York, called Brightside, which contained a large collection of books on typography. During a pleasure trip with his wife and daughter, Richard March Hoe died suddenly in Florence, Italy, on June 7, 1886.

Radio

The first United States patent for a radio circuit was granted to Guglielmo Marconi on July 13, 1897. The patent number was 586,193.

Although Marconi made great advances in the radio field, others who had preceded him had paved the way. These include James Clerk-Maxwell, a Scottish physicist, who in 1873, published a treatise that stated there were electromagnetic wavelengths above and below the wavelength of visible light; Heinrich Hertz, a German physicist, who in 1887, demonstrated by the use of oscillatory radiating circuits that invisible electromagnetic waves travel in air at the speed of light; Edouard Branly, a French physicist, who in 1891, invented the coherer, a long tube filled with iron filings, the electrical resistance of which increases when struck by radio waves; and David Hughes and Alexander Lodge of England and Alexander Popov of Russia who experimented with wireless telegraphy.

Guglielmo Marconi was born April 25, 1874, in Palazzo Marescalchi, the family's townhouse in Bologna, Italy. His father, Giuseppe Marconi, was a wealthy estate owner of the Villa Griffone at Pontecchio, Italy. Young Guglielmo's early education was taken care of by his mother and private tutors, including Augusto Righi. He later attended the Institute Cavellero at Florence and the Leghorn Technical Institute at Livorno, where he developed an interest in chemistry and physics.

In 1894, Marconi read an article in an electrical journal about Heinrich Hertz and his discovery that electromagnetic waves travel through space at the speed of light. It was then that he decided to devote some time to the study of wireless telegraphy. Being born to a wealthy family, Marconi had no need to earn money, so he could devote his time to do anything he pleased. He was given two rooms at the Ville Griffone at Pontecchio for his wireless experiments.

It was there that he assembled the Hertz transmitter, the Branly coherer, and the Popov antenna, and began to experiment with wireless transmission. After improving these components, and with the help of his brother Alfonso, he was able to transmit and receive Morse code from 30 feet across the room to a mile across the garden in 1896.

To satisfy his three-year military service requirement, Marconi was assigned as a naval student to the Italian Embassy in London in 1896. The first

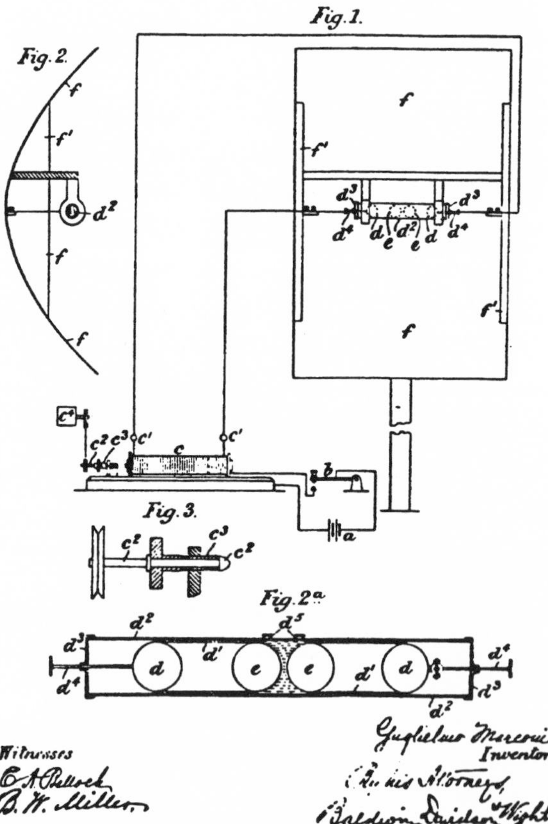

Guglielmo Marconi's work in the radio field led to this patent for radio circuitry, granted on July 13, 1897

thing he did on arriving in London was to apply for a patent on his wireless invention. His cousin, Henry James Davis, prepared the patent application. Marconi was granted British Patent No. 12,039 on June 2, 1896. The equivalent United States Patent, No. 586,193, was issued to him on July 13, 1897 and was reissued on June 4, 1901 as United States Reissue No. 11,913.

Marconi met William Preece, engineer-in-chief of the London General Post Office, who helped him demonstrate his apparatus from the roof of the post office building to a house on the Thames Embankment. Preece then helped him install the first wireless telegraph station at Lavernock Point, near Cardiff, where signals were sent and received in May 1897 on the Island of Flat Holme in the middle of the Bristol Channel, a distance of about three miles. Marconi was assisted in this work by James S. Kemp, who remained with him until his death.

Within a short time, the story of Marconi's success was known all over Europe. He and a small group of investors formed the Marconi Wireless Telegraph Company in London in July 1897 and the Marconi Wireless Telegraph Company of America in 1899, which became a part of RCA in 1919.

By the end of 1897, Marconi and Kemp were sending signals a distance of 10 miles and by the end of 1900, the distance had increased to 150 miles. They accomplished this by adding a condenser and tuning coils to the transmitter and by adding tuning coils to the receiver so that the two circuits were tuned to each other.

On December 12, 1901, Marconi and Kemp, using an antenna attached to a kite at St. John's, Newfoundland, heard signals from associates, headed by J. A. Fleming, in Poldhu, Cornwall, England, a distance of 2,000 miles. This marked the beginning of transatlantic messages with the first west-to-east message being sent from Glace Bay, Nova Scotia, to England on December 17, 1902. The first exchange of greetings from heads of state were between President Theodore Roosevelt and King Edward VII from station WCC at South Wellfleet, Cape Cod, Massachusetts, on January 19, 1903. Commercial service began October 17, 1907, between Glace Bay and Clifden, Ireland.

During World War I, Marconi was made Commander of the Italian Navy. In 1919, he was named the Italian delegate to the Paris Peace Conference, signing for Italy the peace treaties with Austria and Bulgaria.

In addition to receiving more than 100 patents relating to wireless communication, Marconi received many honors, including sharing the Nobel Prize for physics with Ferdinand Braun in 1909, the Franklin Medal of the Franklin Institute, the Albert Medal of the Royal Society of Arts in London, and from the Italian government the hereditary title of marchese, an appointment to the Senate, and the Grand Cross of the Order of the Crown of Italy.

Radio signaled the message around the world on July 20, 1937, that Guglielmo Marconi had died in Rome.

Reaper

The invention of the reaper was a major contributing factor to the prosperity of the United States. The first U.S. patent for a commercially successful reaper was granted to Cyrus Hall McCormick on June 31, 1834.

McCormick was not the first to build a reaper, but he was the first, through his extraordinary business ability, to introduce the reaper to many American and European farmers. In 1800, Joseph Boyce of England received the first patent for a reaper. Beginning in 1803, over 20 U.S. patents were issued for reapers before the McCormick patent. Included among these patentees were Richard French and John F. Hawkins of New Jersey, in 1803; James Ten Eyck of Bridgewater, New Jersey, in 1825; William Manning of Plainfield, New Jersey, in 1831; T. D. Burrall of New York, in 1832; William and Thomas Schnebly of Maryland, in 1833; Abram Randall of Oneida County, New York, in 1833; and Obed Hussey of Nantucket, Massachusetts, in 1833. In Europe, there were Robert Salmon of Woburn, England, in 1807; Smith of Deanston in Perthshire, in 1811; Henry Ogle of Remington, England, in 1822; and the Reverend Patrick Bell of Forfarshire, Scotland, in 1826. However, none of these reapers, with the possible exception of the Hussey reaper, were commercially successful.

Cyrus Hall McCormick was born February 15, 1809, on Walnut Grove Farm, Rockbridge County, Virginia. His formal schooling was very limited and consisted of only the local school when he wasn't needed on the farm. His father, Robert McCormick, had on his 1,600-acre farm a blacksmith shop, a saw mill, a grist mill, a distillery, and a smelting furnace. In addition to being a farmer, he was also an inventor. He had invented a thresher, a bellows, a clover huller, and a hemp binder, and he had worked for several years on a mechanical reaper. As young Cyrus grew older, he often helped his father work on this machine, but it never reached perfection. Finally, Robert abandoned the idea and left the building of a reaper to his son.

Young Cyrus had always shown mechanical ingenuity. At 15, he had made a cradle for a scythe that was light enough for him to use; at 18, he had made a set of surveying instruments; at 22, he had received a patent on a hillside plow; and at 23, he patented a self-sharpening horizontal plow. When he started to work on a reaper, it took him only six weeks

C. H. McCORMICK.
Reaper.
Patented June 21, 1834.

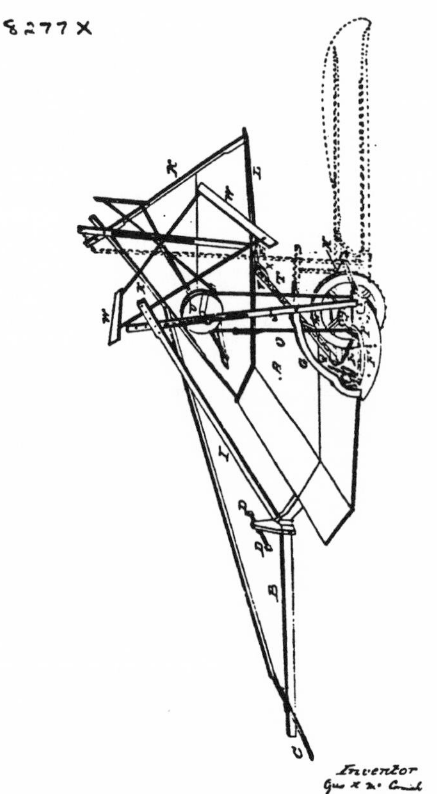

The reaper was developed and introduced to American and European farmers by Cyrus Hall McCormick

to develop a workable one that proved successful in a public trial on the farm of John Steele at Steele's Tavern, Virginia. A few days later, after making some improvements, he had another successful exhibition of his reaper on the farm of John Roff, near Lexington, Virginia. McCormick first advertised his reaper in the *Lexington Union* on September 14, 1833 at $50 each, but he had no buyers.

For a while after this test, McCormick set the reaper aside to make some of his hillside plows which were in great demand. In fact, he did not even think of taking out a patent on his reaper until he read a notice published in the April 1834 *Mechanics' Magazine* of New York City that told of a reaper invented by Obed Hussey and patented in 1833. This moved him to take out a patent on June 21, 1834. For the next few years, he spent some time improving his reaper, but most of his time was consumed in making pig iron at the Cotopaxi iron furnace, together with his father. This business prospered until the panic of 1837 which wiped out the McCormick iron works and left them in debt. It was then that Cyrus began to concentrate seriously on the commercial manufacture of his reaper.

The construction of the reaper first took place in the blacksmith shop at Walnut Grove, Virginia. In 1840, McCormick sold one; in 1842, he sold seven reapers; in 1843, 29; and in 1844, 50, all for $100 each. In 1844, McCormick made a trip through the northern and midwestern states and saw the potential of his reaper on the vast flatlands. Returning home, he made some further improvements, and took out a second patent on the reaper, United States Patent No. 3,895, on January 31, 1845. He then licensed others to manufacture his reapers at Brockport, New York, and Cincinnati, Ohio. However, these licensees produced machines of inferior materials and workmanship. It was then that McCormick decided to build his machines in one place under his own direction.

In 1847, he built his own factory in Chicago, Illinois, in partnership with William B. Ogden, the first mayor of Chicago. Also in 1847, he took out a third patent for an improved reaper, United States Patent No. 5,335, on October 23, 1847. By 1849, 1,000 reapers had been manufactured and sold from the Chicago factory. He had made enough profit to be able to buy out Ogden's share in the business and admit his two brothers into partnership. McCormick introduced or adopted several business practices that were rare or unknown at the time. He set up a rigid inspection system, used standardized parts, gave a written guarantee, sold at a fixed price, used direct-mail advertising, appointed district agents, built storage warehouses in each district, instituted time payments, conducted field tests, and gathered testimonials from prominent figures. These business practices enabled him to out distance his competitors, of which there were many. In 1847, McCormick and Hussey were the only manufacturers. By 1850, there were at least 30 rivals and by 1860, more than 100. But Hussey was McCormick's greatest competitor, and for years both men manufactured reapers and battled each other, not only in the field but also in court. Finally, in 1858, Hussey, unable to compete with McCormick's business abilities, sold out to another of McCormick's rivals, William F. Ketchum, for $200,000, and retired.

In 1851, after his reaper was well known in the United States, McCormick turned his attention toward Europe. He introduced his reaper at the world's fair in London and won the Grand Council Medal. The London *Times* had called his invention "a cross between an Astley chariot, a wheelbarrow and a flying machine," but later conceded that the McCormick reaper was worth the whole cost of the exposition. In 1855, he won the grand prize at the Universal Exposition in Paris. In 1867, Napoleon III, himself, awarded the cross of the Legion of Honor to McCormick. Between 1851 and 1880, McCormick won major prizes at world fairs at Paris, London, Hamburg, Lille, Vienna, Philadelphia, and Melbourne. In 1879, he was elected a member of the French Academy of Sciences which declared that he had done more for the cause of agriculture than any other living man.

No sooner had his reaper become a commercial

success than McCormick found himself entangled in endless litigation. In 1854, McCormick sued the firm of Manny & Emerson for $400,000 for infringing his improvement patents. Manny, with the financial help of all of McCormick's competitors, hired Edwin M. Stanton and Abraham Lincoln to represent him in court. The case was fought all the way to the Supreme Court, where McCormick lost. In 1862, McCormick sued the Pennsylvania Railroad for losing his luggage. The case went to the Supreme Court five times. McCormick finally prevailed, but died before payment was received. The railroad paid his estate $18,060, the value of the luggage plus interest, but it was not nearly enough to cover the litigation costs.

The reaper made McCormick a multimillionaire, but he was generous with his money. In 1859, he endowed four professorships in the Presbyterian Theological Seminary of the Northwest, which in 1886 was named McCormick Theological Seminary. He also made generous contributions to the Union Theological Seminary at Hampden-Sidney, Virginia, and the Washington College at Lexington, Virginia.

McCormick died May 13, 1884, in Chicago. His son Cyrus, together with George W. Perkins, engineered a merger of six of the largest makers of agricultural machinery, including the McCormick Company, to form the International Harvester Company.

Revolver

The first practical and commercially successful revolver was invented by Samuel Colt, who received a United States patent on February 25, 1836. The patent was reissued as United States Reissue No. 124 on October 24, 1848. He also patented improvements on this revolver and was granted United States Patent Nos. 1,304 on August 29, 1839, 7,613 on September 3, 1850, and 7,629 on September 10, 1850.

A revolver is a firearm that consists of a fixed barrel, a firing mechanism, and a revolving cylinder. If it is small enough to be held and fired with one hand, it is classified as a pistol. In 1540, Caminelleo Vitelli of Pistola, Italy, invented the first handheld firearm and named it a "pistol" after the name of the city. It was not a revolver since it did not have a cylinder and fired only one shot at a time. In 1718, James Puckle, a London lawyer, was awarded British Patent No. 418 of 1718 that describes the first revolver mounted on a tripod. A chamber held six shots that when fired could be replaced by a full chamber. In 1813, Elisha Collier of Boston, Massachusetts, invented the first five-shot flintlock chamber that consisted of an elongated cylinder having a number of bores. When the trigger was pulled, the cylinder rotated and at the same time cocked the hammer and fired it. Collier, finding no interest in America, went to England, where his revolver was manufactured. One of these found its way to Calcutta, India, where Samuel Colt first saw it. The first revolver patented in the United States was issued to D. G. Colburn on June 29, 1833. This gun was never publicly accepted. It was Samuel Colt who first designed and patented a revolver that achieved widespread acceptance and use in the United States and abroad.

Samuel Colt was born July 19, 1814, in Hartford, Connecticut. His father Christopher Colt was a manufacturer of cotton and woolen fabrics. His mother Sarah, the daughter of Hartford's first insurance company president, died when Samuel was six years of age. His father remarried within two years, and as a result of family friction, young Samuel lived for a while with his paternal aunt. In 1824, his father sent him to work in his dyeing and bleaching factory in Ware, Massachusetts, where in addition to working in the factory and on nearby farms, Colt also attended the village school. In 1828, his father sent him to Amherst Academy in Amherst, Massachusetts. He remained there until July 4, 1830, when a pyrotechnic display that he participated in burned some school property and caused him to withdraw from the academy.

Young Colt persuaded his father to let him go to sea, and on August 2, 1830, he sailed from Boston on the ship *Corlo* as a common sailor, bound for Calcutta, India. Colt had always had a love for firearms,

4 Sheets—Sheet 1.

S. COLT.
Revolving Gun.

No. 124.

Reissue Oct. 24, 1848.

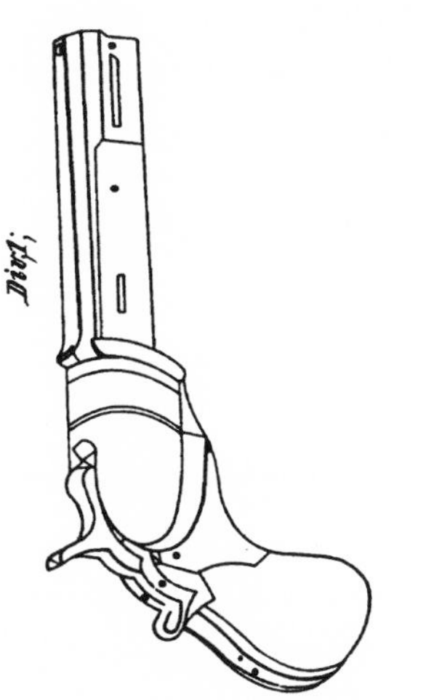

Samuel Colt's Revolving Gun, as reissued on October 24, 1848

having taken apart and reassembled many guns of neighbors while working in Ware. This love was rekindled when he saw the Collier revolver in Calcutta. He determined to improve this gun, and during his spare time on the return trip, he whittled a wooden model of his revolver out of white pine, using a chisel and a jackknife.

Upon his return to America in 1831, Colt engaged the services of Anson Chase, a Hartford gunsmith, to build samples of his revolver. Colt, impatient for success, tried one of these samples before it was perfected. It blew to pieces in his hand. Without money or employment, he returned to work in his father's dyeing and bleaching factory in Ware, temporarily abandoning the revolver business. It was there that he learned about nitrous oxide from the head chemist of the bleaching department. Perceiving that money could be made from public demonstrations of the effects of laughing gas, he set out as an entertainer in 1832 under the name of "Dr. Coult." He visited many large cities in the United States and Canada. It was a profitable venture, and with the first profits he again engaged Chase to work on a revolving rifle. At the same time, he employed John Pierson and Frederick Hanson, machinists of Baltimore, Maryland, to make by hand two pistols in which the cylinder was automatically locked in position at the instant of firing. One of these pistols was stocked and engraved by Richard B. Henshaw of New York City and was used as a specimen and working model in Colt's attempt to form a manufacturing stock company.

On October 22, 1835, England was the first country to grant Colt a patent on his revolver. The United States followed on February 25, 1836. Armed with these patents, Colt persuaded a group of investors to back him in forming the Patent Arms Manufacturing Company in Paterson, New Jersey, and by the summer of 1836 his revolvers were being made.

The revolvers were not too popular at first. Most of the produced revolvers were sold in Texas, which in April 1836 had won independence from Mexico. The government was not interested but did purchase a few because of the Seminole Indian War in the Everglades of Florida. In 1841, with little demand for firearms of any kind, the Patent Arms Manufacturing Company failed.

Colt then turned his attention to protecting American harbors with submarine mines that exploded under water by electrical impulses. He also manufactured waterproof telegraph cable, organized a telegraph company, and installed telegraph lines under a royalty arrangement with Samuel Morse. Then came the Mexican War in January 1846. The government ordered 1,000 revolvers from Colt, but he had no means of making revolvers in such volume at this time. He sought aid from Eli Whitney Jr., who made them for him at the Whitney Arms Company in Whitneyville, Connecticut. In November 1847, an order from the government for a second 1,000 revolvers was received. Colt manufactured these in a small rented building on Pearl Street in Hartford, and hired Elisha K. Root as manager. In late 1847, Colt received a third order for 1,000 revolvers from the government.

With an able man in charge and money for travel, Colt went to Europe to persuade the rulers of the various countries to buy his revolver. The trip was successful and resulted in many orders, including one for 5,000 from the Sultan of Turkey. New factory space was needed for the increased business, so another plant was built between Grove and Potter Streets in Hartford. Colt's Patent Firearms Manufacturing Company was then organized.

In February of 1849, Congress granted a seven-year extension to Colt's revolver patent. In the same year, a United States patent was granted to Edwin Wesson for a similar revolver. Colt's attorneys instituted an infringement suit against the Massachusetts Arms Company that was making revolvers under the Wesson patent. The final verdict in favor of Colt ensured his monopoly of the revolver trade until 1856.

In 1852, Colt bought 250 acres of Hartford land on the west side of the Connecticut River and began to build a new $2 million factory, which was finished in 1855. In the meantime, he had opened a factory in London that operated from 1851 until 1855. The Crimean War in Europe and the Civil War in the United States caused Colt to prosper even more. From 1861 to 1865, Colt's company made and sold to the United States government alone 387,017 revolvers, 113,980 muskets and 7,000 rifles.

In 1860, Colt developed a rheumatic condition that first crippled him and then killed him. He died on January 10, 1862, in Hartford, the same town in which he was born. He left his widow a huge fortune, large sums of which she donated to charity.

Rocket

The first United States patent for a practical multistage rocket was awarded to Robert Hutchings Goddard on July 7, 1914. The number was 1,102,653.

Although Goddard is described as the "father of the modern rocket," there were others whose contributions also played a part in the development of the rocket. The Chinese developed and used rockets in the 11th century. In 1800, Sir William Congreve of England did a systematic research on the improvement of solid-fuel rockets for military purposes. Some of his rockets weighed up to forty pounds each and had a range of more than one mile. The Congreve rockets were used in the War of 1812 at Fort McHenry in Baltimore and inspired Francis Scott Key to use the phrase "the rockets' red glare" in the *Star Spangled Banner.*

On June 21, 1859, Andrew Lanergan of Boston, Massachusetts, was awarded the first United States patent, No. 24,408, on rockets. Thirty-three other patents had been granted for improvements to rockets by the United States before the Goddard patent.

In 1903, Konstantin E. Zoilkovsky, a blind Russian schoolmaster at Kaluga, published the first scientific writing in *The Scientific Observer* on the subject of rockets. He worked out mathematically many of the requirements for space travel. He also provided data on liquid propellants for rocket propulsion. In 1907, Robert Esnault-Pelterie, a French engineer, completed calculations to show the possibility of high-speed rocket transportation. In 1923, Hermann Oberth, a Rumanian mathematician, published a book, *The Rocket Into Interplanetary Space*, which contained a theory of space travel as well as detailed equipment designs. However, it was Robert Hutchings Goddard who paved the way for rocket propulsion in space exploration.

Robert Hitchings Goddard was born October 5, 1882, in Worcester, Massachusetts. Shortly thereafter, his family moved to the Roxbury section of Boston, where his father, Nahum Danford Goddard, was part owner of a shop that manufactured machine knives used in cutting paper and wood. It was in Boston that young Goddard received his early education. Reading a copy of H. G. Wells's *The War of the Worlds* in 1898 kindled his imagination for space travel, a quest that he pursued for the rest of his life.

In 1898, the family moved back to Worcester, where Robert graduated from South High School in 1904. He enrolled at Worcester Polytechnic Institute, graduating with a master's degree in 1910. He completed his work for a PhD at Clark University in Worcester in 1912, and accepted a research fellowship at Princeton's Palmer Physical Laboratory.

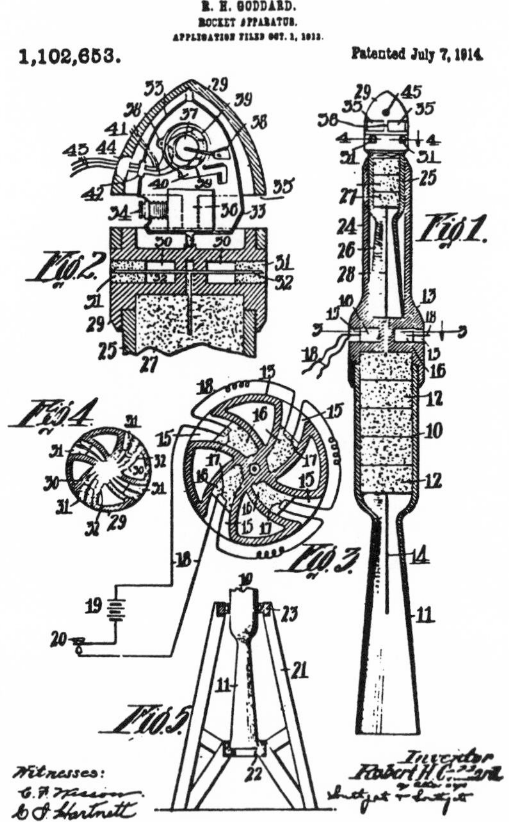

Robert Goddard's multistage rocket

During his college and graduate studies in physics, Goddard maintained his interest in space travel, especially rocket propulsion.

While at Princeton, he was diagnosed as having tuberculosis and was forced to return to Worcester in the spring of 1913. During his convalescence, he prepared two patent applications on rocket propulsion that matured into United States Patent Nos. 1,102,653 on July 7, 1914 and 1,103,503 on July 14, 1914. These basic patents covered the combustion chamber and nozzle, the mechanism of feeding liquid or solid propellant into the combustion chamber, and multiple-stage rockets.

Goddard secured a part-time teaching position at Clark University in 1914, which allowed him time for rocket experiments that he financed out of his own meager salary. In 1916, after exhausting his own resources, he applied for a grant to the Smithsonian Institution, enclosing a 69 page manuscript on the theory of rocket propulsion entitled *A Method of Reaching Extreme Altitudes*. The Smithsonian not only awarded him a $5,000 grant but published his paper in 1919.

Goddard continued his rocketry work until World War I, when he volunteered his services. The Army Signal Corps used his talents to work on the military possibilities of rockets at the Mount Wilson Solar Observatory. There, with his associate, Clarence N. Hickman, he developed a recoilless infantry rocket light enough for an infantryman to carry. The Armistice placed this weapon on the back burner; it did not surface again until World War II, when it became known as the bazooka.

Following World War I, Goddard returned to Clark University and became head of the physics department in 1923. It was about this time that his focus changed from solid to liquid-fuel rockets. On March 16, 1926, he fired the world's first liquid-fuel rocket at Auburn, Massachusetts. It was 10 feet long and was powered with liquid oxygen and gasoline. It rose 41 feet, traveled laterally 184 feet, and its flight lasted 2 1/2 seconds. The flight was witnessed by Henry Sachs, a machinist and instrument maker of Clark University, Dr. P. M. Roope, assistant professor of physics at Clark, and Mrs. Goddard, who made pictures of the event.

Goddard continued his research on liquid-fuel rockets during 1927 and 1928, and on July 17, 1929, he tested an 11 1/2-foot model carrying a barometer, a thermometer, and a small camera. Although the rocket reached a higher altitude than any achieved in previous attempts, it crashed in flames, causing the state fire marshall to ban any further testings.

The press account of the July 17, 1929, noisy, fiery test caught the attention of Charles A. Lindbergh, the noted aviator, who persuaded Daniel Guggenheim of the Daniel and Florence Guggenheim Foundation to provide a $50,000 grant to Goddard for continued research in the rocket field. Goddard took a leave of absence from Clark University and moved his staff to the Mescalero Ranch near Roswell, New Mexico, in the summer of 1930. During the next 11 years, Goddard and his staff succeeded in developing rockets that reached an altitude of more than 9,000 feet and a speed greater than that of sound.

World War II brought an end to the New Mexico rocket experiments, when Goddard again offered his talents to the Armed Services. In 1941, the navy asked him to develop a liquid-fuel, variable-thrust jet for aircraft and transferred him to the Naval Engineering Experiment Station at Annapolis, Maryland. The military rocket development was transferred to the Jet Propulsion Laboratory at the California Institute of Technology. Goddard remained at Annapolis, but his health began to decline. He died at the University of Maryland Hospital in Baltimore of throat cancer on August 10, 1945, and was buried at Hope Cemetery in Worcester, Massachusetts.

Goddard's work is exemplified in his historic paper of 1919 and the awarding of 214 United States Patents. In 1960, the U.S. government acquired rights to the Goddard patents for $1 million by settling an infringement suit filed by the Guggenheim Foundation and Mrs. Goddard. In 1961, the Goddard Space Flight Center in Greenbelt, Maryland, was named in his honor.

Roll-Film Camera

The first United States patent, No. 388,850, for a roll-film camera was awarded to George Eastman on September 4, 1888. This simple camera, together with the roll-film, ushered in the era of popular snapshot photography.

The first optical device that projected scenes through a small opening onto a flat surface so that they could be drawn was called the camera obscura and was devised by 15th-century artists. It was usually a darkened chamber or room having an aperture through which the reflected light from an object passed and produced an inverted exact image of the object. Improvements to the camera obscura consisted of lenses to sharpen the image and mirrors to convert the inverted image. In 1660, it was reduced in size to a box about two feet long.

To permanently capture the camera obscura images, scientists began to experiment with various chemicals to accomplish this purpose. In 1727, Johann Heinrich Schulze, a German physician, observed that silver nitrate darkened when exposed to sunlight but that the images were not permanent. In 1826, Joseph Nicephore Niepce, a French physicist, produced a permanent image using bitumen and washing out the unexposed areas. In 1835, Louis J. M. Daguerre, a French inventor and a partner of Niepce for several years, reduced the time of exposure of iodized silver by developing the latent image with mercury vapor. In 1839, he was able to fix the image by washing out the unexposed silver salts with sodium hyposulfite. In 1841, William Henry Fox Talbot, an English scientist and mathematician, patented the first silver salt negative-positive process for making photographs. In 1851, Frederick Scott Archer, an English sculptor, was the first to use collodion on a glass plate as a binder for light-sensitive silver salts. These plates had to be exposed and developed while wet. In 1871, Richard L. Maddox, a British physician, replaced collodion with gelatin which resulted in the dry-plate process. The light-sensitive plates could now be exposed and developed in the dry state so photographers no longer had to carry around a darkroom and chemicals to make a photograph. However, the person who in 1888 revolutionized the photographic field by introducing a small handheld camera with an integral roll-film holder was George Eastman, who called his camera a Kodak, a word that he coined for use as a trademark.

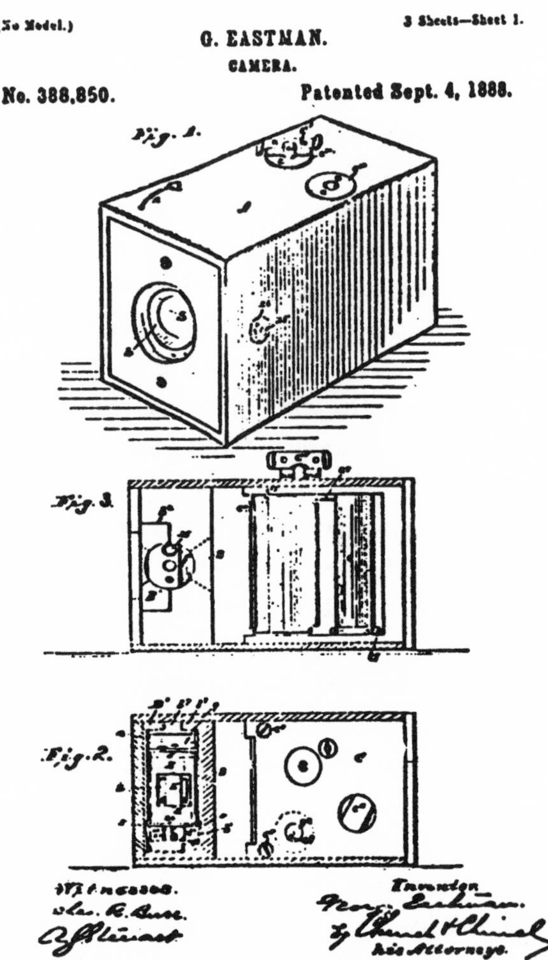

George Eastman's box camera ushered in the era of popular snapshot photography

George Eastman was born on July 12, 1854, in Waterville, New York. The family moved to Rochester, New York, in 1860. His father, George Washington Eastman, was a teacher of penmanship and established the first commercial college in Rochester. Young George's father died when the boy was only eight years of age. His public schooling was limited to only seven years before he took a job in an insurance office at $3 a week to help support his family. In 1874, he transferred to the Rochester Savings Bank as a junior bookkeeper.

Eastman's job at the bank not only enabled him to provide some financial relief for the family but also provided him with the means to pursue his hobby of photography. He was so intrigued by the subject that he read all the available literature on photography. In 1877, his interest was captured by a formula in an English almanac for a dry-plate emulsion. He began experimenting with his own emulsions in his mother's kitchen at night after working in the bank during the day. In 1879, he invented a machine for coating his emulsions on glass plates, for which he received patents in England and the United States.

In 1880, Eastman formed a partnership with Henry A. Strong, leased the third floor of a building on State Street in Rochester, and went into the business of manufacturing dry plates, making arrangements with A & H T Anthony of New York to sell the plates to the photographic trade. In 1881, Eastman resigned his position at the bank and established the Eastman Dry Plate Company. A shipment of plates that lost their sensitivity almost caused the company to collapse. Eastman shut down the operation, and with Strong sailed to England in an attempt to discover what had gone wrong. They discovered that the problem was not in the emulsion formula but in a bad batch of gelatin that had been used. They came back to Rochester, reopened their factory, and shipped new plates to replace the inferior ones.

Eastman now embarked on a project for eliminating the heavy and bulky glass plate support. He succeeded in 1884 by coating a paper strip with a solution of collodion and then with the photographic emulsion, for which he received U.S. Patent No. 306,594 on October 14, 1884. He found that he could wind this paper strip with its emulsion into a roll. Eastman and William H. Walker, an associate, developed in 1885 a mahogany roll holder to fit any standard plate camera. Also in 1885, Eastman marketed a stripping film called Eastman American film. The paper on this film was stripped off after development, leaving a thin transparent negative that was used in making prints.

Eastman began to think in terms of allowing the public to enjoy the results of photography instead of only a few professional photographers. In 1888, he placed on the market at $25 a small box camera containing a roll of film sufficient for 100 exposures. The user returned the camera and exposed film, along with $10, to Rochester, where the film was developed, prints made, and the camera reloaded and shipped back to the amateur. This camera was called the Number One Kodak Camera. Eastman created an advertising slogan for the camera: "You press the button—we do the rest."

Eastman was not satisfied with his paper-backed film because it had a tendency to curl when wet. He became one of the first industrialists to employ a research scientist when he hired Henry M. Reichenbach, a chemist, to develop a flexible and transparent film. In 1889, Reichenbach was successful in producing such a film and applied for a patent, but not before Hannibal Goodwin, an Episcopal minister, had applied for a patent on the same invention two years earlier. A bitter legal battle ensued for many years, including not only Eastman but Goodwin's widow and the Anthony and Scovill Company which had purchased a part interest in the Goodwin patent and was now producing the Goodwin product as Ansco film. The case was settled in 1914. Goodwin was acknowledged as having priority and Eastman was forced to pay several million dollars in retroactive royalties.

In 1891, the company became known as the Eastman Company and the following year, as the

Eastman Kodak Company. During the next several years, the company put on the market several products that the public bought with enthusiasm. These products included daylight-loading film in 1891, the Pocket Kodak camera in 1895, X-ray film in 1896, the Folding Pocket Kodak camera in 1898, the Brownie camera in 1900, cellulose acetate safety film in 1908, 16 millimeter reversal movie film in 1923, 16 millimeter Kodacolor movie film in 1928, 8 millimeter film, cameras, and projectors in 1932, as well as many other products.

Eastman's goal of bringing photography within the reach of nearly everyone was realized. The public's acceptance of Kodak products and services caused the company to prosper and made Eastman a very wealthy man. However, he never forgot the employees who made all this possible. He initiated an employee suggestion system in 1898, a bonus award in 1899, an employment stabilization program in 1903, a pension fund in 1911, a wage dividend payment in 1912, and a retirement program in 1928.

During his latter years, Eastman became one of America's noted philanthropists. He gave more than $1 million to art, education, scientific, and medical institutions, much of it during his lifetime. On March 14, 1932, George Eastman took his own life after leaving a note that read, "My work is done, why wait?" In 1949, his Rochester home became a public museum of photography.

Sewing Machine

The Industrial Revolution brought forth many new and useful products but perhaps none as important as the sewing machine. The first United States patent for a practical sewing machine was awarded to Elias Howe Jr., on September 10, 1846. The number was 4,750.

Howe was not the first inventor of the sewing machine nor was the United States the place where this invention took place. Many people contributed to the idea that a machine could sew pieces of fabric together faster and better than the hand.

Charles F. Weisenthal, a German mechanic living in London, was granted British Patent No. 701 on July 24, 1755, for a two-pointed needle with an eye near one of the points to be completely passed through the fabric. This needle was attached to a crude embroidery machine. Thomas Saint, an English cabinetmaker, was issued British Patent No. 1,764 on July 17, 1790, for a machine that produced from a continuous thread a chainstitch and that incorporated many features common to the modern sewing machine. This patent lay overlooked in the British Patent Office until 1873. John Duncan, a Glasgow manufacturer, was granted British Patent No. 2,769 on May 30, 1804, for a machine that made a chainstitch and used several hooked needles that operated simultaneously. The machine enjoyed some success in embroidery work.

Edward and William Chapman were granted British Patent No. 3,078 on October 30, 1807, for a machine that used two pointed needles, each with an eye near the point, one on each side of the fabric, that did not pass completely through the fabric. The disadvantage was that the thread had to be removed from the first needle and threaded into the second needle after each stitch. Balthasar Krems, a hosiery worker of Mayen, Germany, invented in 1810 a sewing machine that used a pointed needle with an eye near the point and a hook-shaped pin on the opposite sides of the fabric to form a chainstitch. This machine was never patented and did not reach commercial success.

Josef Madersperger, a tailor in Vienna, Austria, received an Austrian Patent in 1814 for a sewing machine that stitched straight or curved lines but the machine was not practical. The Reverend John Adam Dodge of Monkton, Vermont, invented in 1818 a sewing machine having a double-pointed needle with an eye in the middle. He secured the help of John

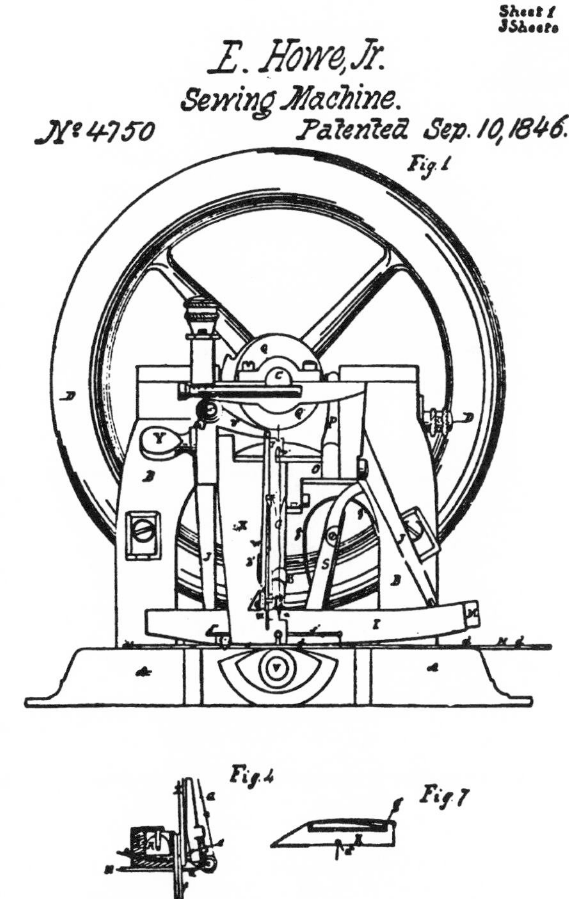

One of the industrial revolution's most useful products, the sewing machine, was patented by Elias Howe Jr., on September 10, 1846

Knowles, a local mechanic, to construct the machine but failed to apply for a patent, and because of the pressing work of his pastorate, he never attempted to manufacture any machines. Henry Lye received a United States patent on March 10, 1826, for a machine for sewing leather. No details of this machine are known because the Patent Office fire of 1836 destroyed the original papers and model, and they were never restored.

The first commercially successful sewing machine was invented by Barthelemy Thimonnier, a French tailor, who received a French patent on July 17, 1830. This machine made a chainstitch by a hooked needle that caught a thread from a carrier on the opposite side of the fabric and pulled it through the fabric to form a loop. A second pass of the needle caused the second loop to become enchained in the first. He made 80 machines for his Paris shop that were used to make army clothing. These machines were destroyed by a mob of tailors who feared that they would rob them of their livelihood.

Walter Hunt, a New York machinist and inventor of the safety pin, made and sold a sewing machine in 1834 that produced a lockstitch by using an eye-pointed needle working in combination with a shuttle that carried a second thread on the opposite side of the fabric. He was working on several other inventions at the time and soon lost interest in his sewing machine. He sold it, together with the right to patent, to George A. Arrowsmith. Neither Hunt nor Arrowsmith bothered to patent it. It was not heard of for 18 years when it surfaced during a patent litigation suit between Isaac Singer and Elias Howe. Singer tried to get the court to declare the Howe patent invalid because of the Hunt machine, but the court held that Hunt had abandoned his invention.

John J. Greenough received United States Patent No. 2,466 on February 21, 1842, for a sewing machine that used a two-pointed needle with an eye in the middle. The needle was passed back and forth through the fabric with a pair of pincers on each side of the seam. Only one machine, the patent model, was made. Leonard Bostwick patented in the United States in 1843, and in England in 1844, a running-stitch machine in which a pair of toothed wheels corrugated the fabric and pushed it onto a stationary needle. Benjamin W. Bean was awarded United States Patent No. 2,982 on March 4, 1843, for a sewing machine that made a running stitch in which the fabric was forced onto and through a threaded needle. This machine had a very limited use.

George H. Corliss, better known as the inventor of the Corliss steam engine, was awarded United States Patent No. 3,389 on December 27, 1843, for a machine for sewing leather that used two two-pointed needles to form the saddler's stitch. Corliss abandoned efforts to manufacture this machine and turned his attention to improving the steam engine. James Rogers was issued United States Patent No. 3,672 on July 22, 1844, for a sewing machine that had very little commercial appeal. John Fisher and James Gibbons received British Patent No. 10,424 on December 7, 1844, for a sewing machine that used an eye-pointed needle and a shuttle to make a two-thread stitch that was not a lockstitch. There is no evidence this machine ever reached a practical stage.

Although many attempts were made, it remained for Elias Howe Jr., to invent a sewing machine that would eventually be accepted by the public. For this, he received the first United States patent for a practical and successful machine.

Elias Howe Jr. was born in Spencer, Massachusetts, on July 9, 1819. His father was a farmer and had on his farm a gristmill, a sawmill, and a place where cotton cards were manufactured. Young Howe's first job at the age of six was sticking wire teeth into strips of leather to form these cotton cards. In the winter he attended the local district school for only a few years. When he was 11, his father loaned him out to a neighboring farmer for whom he did certain chores for his board and keep. Poor health and lameness caused him to return home to work in his father's mills until he was 16.

In 1835, Howe went to Lowell, Massachusetts, and worked in a machine shop that manufactured cotton-spinning machinery. He lost his job during the financial panic of 1837 and went to Boston to seek employment. He became an apprentice of Ari Davis, a maker of scientific apparatus for Harvard University professors. Under the leadership of Davis, Howe became a skilled machinist. It was there that he overheard a customer remark to Davis that if someone would invent a sewing machine the inventor could become independently wealthy. From that moment, the idea of inventing a sewing machine never left his mind.

Howe married at age 20 and soon became the father of several children. His weekly salary of $9 was not enough to care for his family, so his wife helped out by taking in sewing. As he watched her sew, Howe became serious about inventing a machine that would help relieve her of this burden. In 1843, he started to work in earnest on such a machine. Many months of hard work produced his first sewing machine, which turned out to be a failure. He had tried to mechanically copy his wife's hands as she sewed by using a needle pointed at both ends with the eye in the middle.

Remembering the cotton looms he had worked on in Lowell, the idea occurred to him of using two threads and forming the stitch by using a shuttle and a curved needle with an eye near the point. He quit his job with Davis to work full time on his second machine. To do this, he moved his family in with his father in Cambridge. A rough model of wood and wire convinced him that a machine of steel and iron would work, but he had not the means to buy the necessary materials.

He convinced a former schoolmate, George Fisher, to form a partnership. Fisher agreed to board Howe and his family, provide a workshop, and provide money for materials up to $500 in exchange for a half interest in a patent should one be forthcoming. Howe and his family moved into the Fisher house in December 1844. By July 1845, Howe had completed his machine and demonstrated its usefulness by sewing two suits of clothes, one for Fisher and one for himself.

Howe tried to convince the tailors of Boston to accept his sewing machine but without success. He then publicly exhibited his machine at the Quincy Hall Clothing Manufactory and won a contest between his machine and five of the swiftest sewers in the establishment. Even this did not result in a single order for one of his machines.

Howe finished a second machine in the summer of 1846, and with Fisher took it to Washington, where he deposited it in the Patent Office with his application for a patent. The patent was granted September 10, 1846.

With no sales in sight, Fisher, who had advanced about $2,000, became discouraged and dissolved the partnership by selling his interest in the patent to George W. Bliss. Howe and his family then moved back to his father's house in Cambridge.

Thinking that England might be a more fertile field for his invention, Howe sent one of his machines to London with his brother Amasa, who succeeded in selling it for about $1,250 to William Thomas, a maker of corsets and shoes, in Cheapside, London. The sale included the right to patent the machine in England and an oral agreement to pay Howe a royalty on each machine made and sold. Thomas also advanced passage money for Howe and his family so Howe could adapt his machine for sewing leather. Eight months later, Howe and Thomas quarreled, with the result that Howe lost his $15-a-week job. Without money, he was forced to pawn his first machine and patent papers in order to send his family home. Howe remained in England for several months, working at several machinist's jobs and building another sewing machine. Shortly thereafter, with the help of an English friend, Charles Inglis, and the sale of his just completed machine, he set sail for the United States, paying his passage by serving as cook on the sailing vessel.

Arriving in New York in April 1849, Howe soon heard that his wife in Cambridge was very ill. After receiving some money from his father, he traveled to Cambridge, arriving a few days before his wife died. He was so destitute that he was forced to borrow a suit from his brother-in-law to attend the funeral, his own clothes having been lost, together with all his household goods, on a ship that was wrecked off Cape Cod.

During his absence in London, Howe discovered that some manufacturers had begun to make and sell sewing machines based on his design. He then prevailed upon George W. Bliss, who had purchased Fisher's interest, to advance the money necessary to regain his first machine and patent papers from the London pawnshop and to sue these manufacturers for infringement. Bliss agreed after taking a mortgage on the Howe farm as security. These suits lasted from 1849 to 1854, when the final court decided that Howe's patent was valid and infringed and that Howe was due a royalty on every machine that infringed his patent.

Bliss died about this time, and Howe was able to acquire full ownership of his patent. In March of 1861, Howe petitioned and Congress granted a seven-year extension on his patent. It was not long before his income increased from a few hundred dollars a year to more than $200,000, an enormous sum at that time.

During the Civil War, Howe enlisted and served as a private in the 17th Regiment Connecticut Volunteers, refusing a commission. Once when the government was short of money, he advanced the money necessary to pay the entire regiment.

Elias Howe Jr., died October 3, 1867, in Brooklyn, New York, at the home of a son-in-law and was buried in Cambridge, Massachusetts.

Smokeless Gunpowder

The inventor who received the first United States patent, No. 430,212, for smokeless gunpowder was Hiram Stevens Maxim, on June 17, 1890.

Gunpowder, sometimes called black powder, is a mixture of 15 percent charcoal, 10 percent sulfur, and 75 percent saltpeter, or potassium nitrate. The earliest known use of gunpowder was by the Chinese in 1045, but the formula for this material did not reach the western world until 1242 when Roger Bacon of Oxford University in England published a book in which he told how to make it. In 1320, Berthold Schwarz, a German monk, first used gunpowder to explode shells in a firearm he had invented. Gunpowder was used to fire cannons as early as 1346, but it was Captain Thomas Jackson Rodman of the United States Army who in 1857 improved the shape of this product so it could be used more effectively for the cannon. He perfected the prismatic grain in which the gunpowder was made into grains shaped like hexagonal prisms and had several parallel grooves along the outer edge. When these prisms were placed end to end, the grooves made a continuous channel so that the flame burned outward from the grooves and inward from the surface of the prisms, resulting in a slower but more powerful explosion.

One of the early disadvantages of gunpowder was the cloud of smoke produced when it exploded. In 1885, Hiram Stevens and his brother, Hudson Maxim, both worked on the perfection of a smokeless gunpowder, but it was Hiram who perfected the invention and received a patent.

Hiram Stevens Maxim was born February 5, 1840, in Brockway's Mills, near Sangerville, Piscataquis County, Maine. His father, Isaac Weston Maxim, was a farmer, woodturner, and millwright who spent much time working on the invention of an automatic gun and flying machine. After a perfunctory education, young Maxim, at the age of 14, went to work for carriagemaker Daniel Sweat, in East Corinth, Maine. After a short time, Maxim went to Montreal, Quebec, Malone, New York, St. Jean Crysostome, Quebec, and Brasher Falls, New York, working as a carriage painter, a cabinetmaker, a mechanic, and a bartender along the way.

After his experiences in Canada and upstate New York, Maxim worked as a mechanic in the engineering works of his uncle, Levi Stevens, in

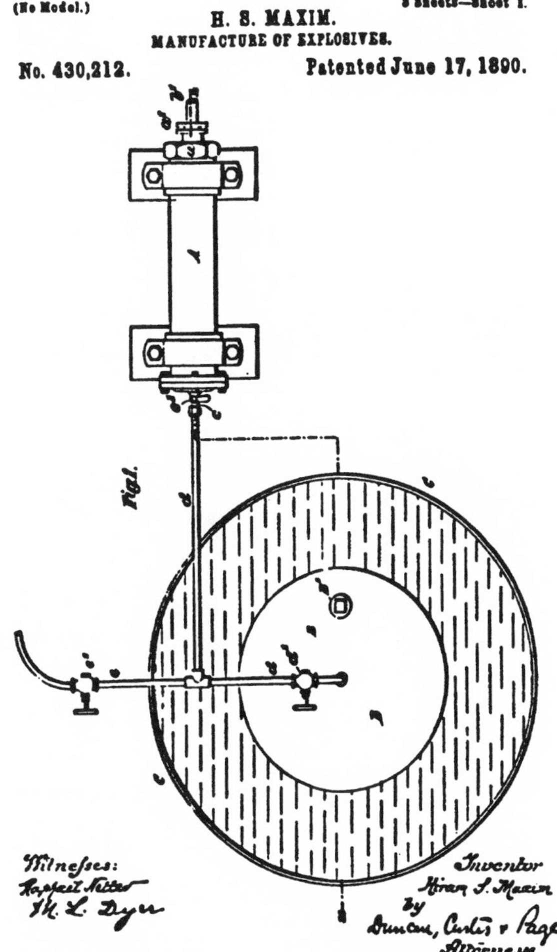

Hiram Stevens Maxim's smokeless gunpowder, patented June 17, 1890

Fitchburg, Massachusetts. In 1866, while working for Oliver P. Drake, a maker of scientific instruments in Boston, he invented and received patents on a hair-curling iron, an automatic repeating mousetrap, and an automatic sprinkler system. In 1870, he moved to New York City to work for the Novelty Iron Works and Shipbuilding Company, where he invented a machine for generating illuminating gas, a locomotive headlight, and a gas engine.

In 1877, Maxim became a consulting engineer and in 1878 became chief engineer of the United States Electric Company, a rival of Edison in the electrical field. Turning his attention to the incandescent carbon lamp, Maxim developed a method of flashing electric light filaments in a hydrocarbon atmosphere to uniformly deposit carbon on the filaments. However, he never received a patent on this, and it became public property both in the United States and England.

In 1881, Maxim went to Paris to show the U.S. Electric Company's equipment at the Paris Exposition. It was there that he conceived an idea for an automatic gun. He traveled to London and set up a laboratory in Hatton Garden to perfect the gun. In 1883, he succeeded in producing a single-barrel machine gun in which the recoil of each shot fired ejected the spent cartridge and moved the next one into the chamber. It fired at a speed of more than 600 shots per minute. The War and Navy Departments of the United States rejected the gun as impractical, but the British government was more receptive. In 1884, Maxim formed the Maxim Gun Company which merged in 1888 with the Nordenfeldt Company. In 1896, the firm was absorbed into Vickers Sons and Maxim; in 1911 the name was shortened to Vickers Ltd. Although his machine gun had been preceded by the Gatling gun of 1862, the mitrailleuse of 1867, and the Nordenfeldt of 1877, his automatic gun of 1883 was the most efficient. He made improvements on his gun and, at the suggestion of Lord Wolseley, invented a smokeless gunpowder of the cordite type.

In 1889, Maxim became interested in flying and in 1894 built a $200,000 steam-driven multiplane. It weighed 8,000 pounds and was driven by a 360-horsepower steam engine. In a trial at Bexley, Kent, the machine actually lifted itself from the ground but crashed shortly thereafter. He never attempted to build another.

Maxim became a British subject in 1900 after living in London since 1883. He was knighted by Queen Victoria in 1901. He received 122 patents in the United States and 149 in Great Britain. His brother, Hudson Maxim, was the inventor of many explosives, and his son, Hiram Percy Maxim, was the inventor of the Maxim silencer, which when applied to gunfire eliminated 95 percent of the noise.

Sir Hiram Stevens Maxim died November 24, 1916, in his home at Streatham, London, England.

Steel Manufacture

The first United States patent for a commercial process for manufacturing steel was awarded to William Kelly on June 23, 1857. The number was 17,628. The first to file for a patent for making steel was Henry Bessemer of England on January 10, 1855, but in an interference proceeding between Bessemer and Kelly, the Patent Office gave priority to William Kelly.

The story of steel is really the story of how to control the carbon content of iron. Iron containing from 0.3 to 1.7 percent carbon is called steel. Anything below this range is called wrought, or malleable, iron and anything above this range is called cast, or pig, iron.

Iron ore is the chief raw material in making steel. Early ironmakers made iron into a usable form by placing the ore in a bed of heated charcoal or coke and allowing the carbon in the charcoal or coke to react with the oxygen in the ore to form a gas that escaped and left iron behind. These early ironmakers were not aware that the carbon content of the iron was responsible for the production of steel. However, some steel in small quantities was made by trial and error before the 18th century.

In 1722, Rene Antoine de Reaumur, a French physicist, made steel in larger quantities by placing wrought iron in a molten bath of cast iron. The next development in the production of steel occurred in 1740 when Benjamin Huntsman, an English instrument maker, improved the homogeneity of steel by discovering the crucible process. The manufacture of steel was better understood in 1750 when T. O. Bergmann, a Swedish metallurgist, discovered the influence of carbon on the hardness of steel. Henry Cort, an English ironmaker, in 1784 advanced the production of steel by building the puddling furnace and the rolling mill. However, the greatest advance in the mass production of steel was the Bessemer process, independently invented by William Kelly in America and Henry Bessemer in England in the 1850s. Although called the Bessemer process because he first brought it to the attention of the world, the first valid U.S. patent was issued to William Kelly in 1857.

William Kelly was born August 21, 1811, in Pittsburgh, Pennsylvania. His father, John Kelly, was a landowner and businessman in Pittsburgh. Young Kelly was educated in the public schools of Pittsburgh and soon developed an interest in metallurgy because the town was becoming important in the manufacture of iron, employing 10 percent of the population for that purpose.

It is not known why Kelly did not pursue his interest in the iron business and instead went into the

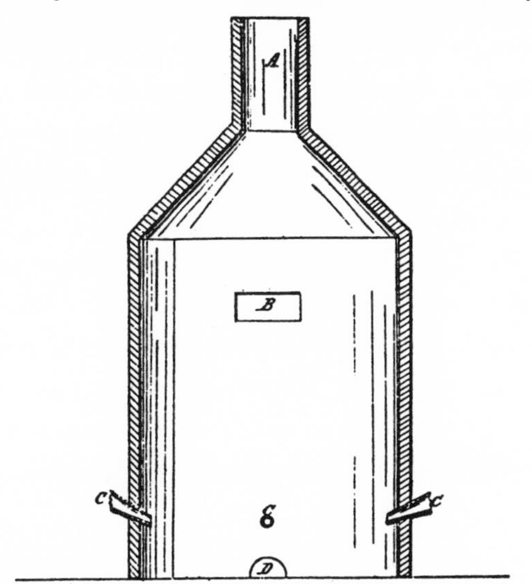

William Kelly's invention of the commercial process for manufacturing iron and steel gave a powerful boost to industrial progress

dry-goods business in Philadelphia with his brother. However, this interest surfaced a few years later when on a trip to Nashville, Tennessee, to collect debts for the business, he met and married a young woman from Eddyville, Kentucky. On visiting Eddyville, he noticed that the land around there was rich in iron ore. He sent for his brother in 1846 and, with a loan from his father-in-law, bought some land and set up the Suwanee Iron Works & Union Forge to manufacture sugar kettles.

The business prospered for a while, but Kelly noticed that the supply of timber for making charcoal was being diminished because the separation of iron from iron ore required large amounts of charcoal. Kelly began to seek a new way to make iron without using too much charcoal. After much experimentation, he observed that if a blast of cold air was introduced into the molten pig iron, wrought iron could be made without the use of any charcoal, the carbon in the pig iron serving as the fuel. He also observed that steel could be produced if the air was shut off at the proper time.

Kelly called his method the pneumatic process. However much he tried, Kelly could not convince his family and customers that he was not crazy and had indeed invented a new way of making wrought iron and steel. Rather than antagonize his family and customers, he continued making iron the old way, but in 1851 he secretly built the first of seven pneumatic furnaces in the woods about three miles away to continue his experiments.

In 1856, on hearing that Henry Bessemer of England had patented the same invention, Kelly applied for a patent on his process. An interference was declared between Bessemer and Kelly, and in April 1857 Acting Commissioner of Patents, S. T. Shugert, decided in favor of Kelly. Bessemer, however, did receive a patent on the tilting converter, a crucible mounted on trunnions. This apparatus was necessary for carrying out the process.

Both Kelly and Bessemer found it difficult to make steel until Robert F. Mushet, a Welsh metallurgist, appeared on the scene. He found that if a small amount of spiegeleisen, an alloy of iron, carbon, and manganese, was added to the wrought iron produced in the Kelly-Bessemer process, the carbon content of iron would be raised to the necessary percentage required for steel. He patented his invention in both England and the United States.

The financial panic of 1857 forced Kelly into bankruptcy. He then went to Johnstown, Pennsylvania, where he convinced Daniel J. Morrell of the Cambria Iron Works to allow him to build a converter to continue his ironmaking experiments. In 1861, Zoheth Durfee and E. B. Ward obtained control of Kelly's patents and with Morrell organized the Kelly Pneumatic Process Company and built a plant at Wyandotte, Michigan. They also obtained the rights to the Mushet patent in the United States. In September 1864, the first Kelly-Bessemer steel made in the United States was made at the Wyandotte plant.

Meanwhile, Alexander Lyman Holley built a Bessemer steel plant in Troy, New York, in 1865 after acquiring the American rights to the Bessemer converter. Immediately, the threat of a litigation surfaced. The Kelly company could not produce steel without the rights to the Bessemer converter, and neither could the Holley interests produce steel without the rights to the Kelly process and the Mushet recarburizing process. In 1866, the problem was solved when the two interests consolidated and formed the Pneumatic Steel Association, a joint stock company. It was not long after this that the United States surpassed England in the production of steel.

Kelly remained at Johnstown for five years, realizing about $450,000 from his invention. He then moved to Louisville, Kentucky, and established an axe-manufacturing business. He died February 11, 1888, in Louisville.

Submarine

The first United States patent for a practical submarine acceptable by the U.S. Navy was granted to John Philip Holland on September 9, 1902. The patent number was 708,553.

Submarine history began with Cornelius Van Drebbel, a Dutch scientist, who invented and built the first submarine in 1620. It was constructed of wood and made watertight by a covering of leather smeared with tallow. It was propelled by oars from the inside, which came out through watertight leather valves or sleeves. It is reported that King James I witnessed some of its trips in the Thames River. It was originally intended as a naval vessel, but the British Admiralty refused to adopt it.

Many other similar underwater vessels were built during the next century, but the next one of historical importance was that built by David Bushnell at Saybrook, Connecticut, in 1775 and named the American *Turtle.* It was built for the purpose of sinking the British ships that threatened New York City during the Revolutionary War. Bushnell's one-man submarine was about six feet in diameter and made of wood. It was powered by two screw propellers, one for depth and the other for lateral motion, both operated by hand. It could submerge by admitting water into a tank by a foot-operated valve and could be made to rise by ejecting the water with two forcepumps. The vessel contained a depth indicator and carried a torpedo that could be attached to the hull of an enemy ship and set to explode by clockwork. It almost succeeded in blowing up the British ship *Eagle* in New York harbor in 1776.

In 1801, Robert Fulton, an American in Paris, built the first cigar-shaped submarine, called the *Nautilus,* from money he received from the sale of his panorama painting "The Burning of Moscow." He tried unsuccessfully to interest the French Republic in his submarine, finally selling his invention to the British government for $75,000. Fulton returned to the United States in 1806 and won lasting fame with his steamboat. In 1814, he built another submarine called the *Mute.* This was a large boat more than 80 feet long, 21 feet wide and capable of carrying 100 men. Before its trial, Fulton died and no one seemed interested in continuing his submarine work.

Wilhelm S. V. Bauer, a German woodturner, built and launched a submarine at Kiel in 1851. After being turned down by the governments of Germany, Austria, and England, Bauer went to Russia. There in 1855, he built a boat called *Le Diable Marin,* which

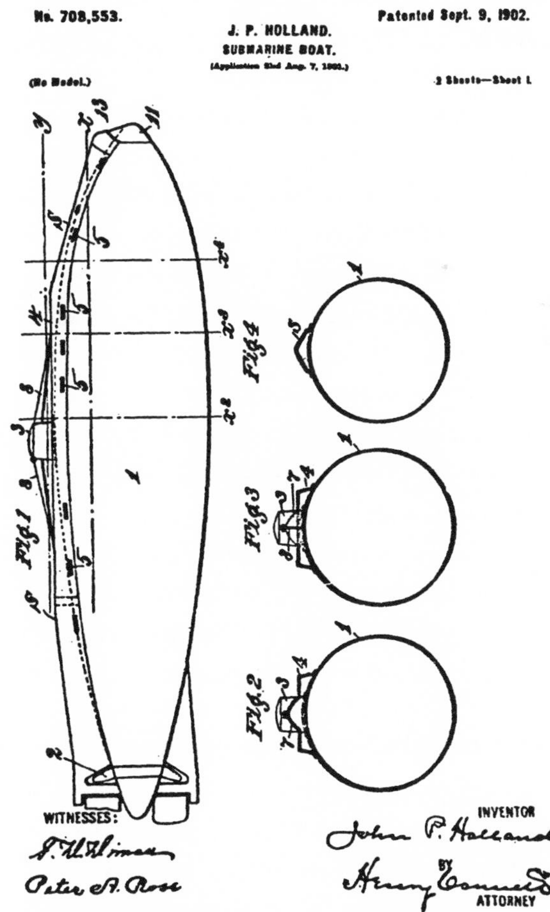

John Philip Holland's Submarine Boat, patented on September 9, 1902

was accepted by the Russian Navy. It was 52 feet in length, 12 feet in width and it made more than 130 successful trials. However, the Russians soon lost interest, and Bauer returned to Germany and his wood lathe in 1858.

During the Civil War, the first enemy ship sent to the bottom was sunk by a torpedo fired from a submarine. The Union warship *U.S.S. Housatonic* was sunk by the submarine *Hunley* in 1864. This sub was built by Captain H. L. Hunley and Lieutenant J. A. Alexander of the Confederate States Army in Mobile, Alabama, in 1862.

Many other workers advanced the art of submarine building. These include O. S. Halstead in 1866 of Newark, New Jersey. Halstead's submarine could be converted to a diving bell when it was resting on the bottom. G. W. Garrett, of Liverpool, England, built a submarine in 1878 that could maintain equilibrium by forcing water in and out of a cylinder by means of a piston. Claude Goubet, a French engineer, introduced electricity as the power source in 1881. George C. Baker of the United States built a vessel in 1882 that used side propellers that could be turned to any desired angle. Thorsten Nordenfelt of Sweden incorporated torpedo tubes within the hull in 1886. Gustave Zede of France built several boats in 1888 which were driven by electricity. Depuy de Lome of France in 1888 was the first to use a gasoline motor on the surface and batteries when submerged. Lieutenant Isaac Peral of Spain built a submarine in 1889 that had an automatic depth finder, a range finder, and a caustic soda device for eliminating carbon dioxide.

However, perhaps no work was more important than that of Simon Lake, who with Holland has been closely associated with the history of submarines. Lake's submarine was fitted with wheels that allowed it to travel on the bottom of the sea. The submarine also had a door that could be opened under water to allow the crew to leave and enter in diving suits. His vessel had a double hull and a series of horizontal rudders to maintain it on an even keel. Despite the several advantages of the Lake submarine, the United States Navy decided on the Holland vessel.

John Philip Holland was born on February 29, 1840, in Liscanor, Ireland. He received his early education in Liscanor and then attended the Christian Brothers schools at Ennistymon and Limerick. He taught school in various places in Ireland from 1858 to 1872. During this time, he became a patriot for Irish independence and in 1870 conceived the idea of a submarine as a weapon to be used against the British Navy. Lacking construction means, he abandoned the project, but the idea of an underwater vessel would remain with him.

Holland came to the United States in 1873 and secured a job as a teacher in St. John's Parochial School in Paterson, New Jersey. After revising his submarine design, he submitted it to the United States Navy in 1875 only to have it rejected. About this time, the Fenian Society, the Irish Republican Brotherhood, supplied Holland with money to build a submarine capable of crossing the Atlantic and destroying the British fleet. Holland built the *Fenian Ram* that was launched in the Hudson River in 1881. This sub was 31 feet long, six feet wide, and had a one-cylinder Brayton internal combustion engine. It contained all the chief principles of modern submarines in balance and control. However, it never accomplished its war purpose and is now located as a memorial in a city park in Paterson, New Jersey.

Holland's reputation as a submarine engineer attracted the attention of the Cramps, famous Philadelphia shipbuilders. They asked him in 1888 to submit plans for a submarine to be built by the United States government. Holland and Thorsten Nordenfelt, the Scandinavian, submitted plans. Secretary of the Navy William C. Whitney selected Holland's design. However, governmental red tape and a change in administration resulted in the project being abandoned.

In 1893, Congress again became interested in submarines and appropriated $200,000 for the construction of an experimental vessel. Inventors were invited to submit plans. Only two other men besides Holland entered the competition: George C. Baker and Simon Lake. In 1895, Navy Secretary Hilary A. Herbert awarded a contract to the J. P. Holland Torpedo Boat Company to build a submarine according to navy specifications. It was built at the Columbian Iron Works, Baltimore, Maryland. Most of Holland's ideas were ignored and replaced by the acts of Admiral George Melville, Chief of the Naval Bureau of Steam Engineering. The completed vessel, called the *Plunger,* did not perform properly and was abandoned as a failure.

With his own money and incorporating his own ideas, Holland began to build another submarine, called the *Holland,* at the Crescent Shipyards, Elizabeth, New Jersey. It was launched in 1898. It had a cruising radius of 1,500 miles on the surface and 50 miles under water and was powered by a gasoline engine for surface cruising and electric storage batteries when submerged. It had the ability to dive by inclining its axis to the horizontal and traveling to the desired depth. On April 11, 1900, after many successful tests, the *Holland* was purchased by the federal government for $150,000 as the navy's first submarine. A few months later six more vessels like it were ordered to be built by Holland. He also built submarines for Great Britain, Russia, and Japan.

Holland retired from his company in 1904 and devoted his final years to aeronautical experiments. He died August 12, 1914, in Newark, New Jersey.

Telegraph

The telegraph was invented by Samuel Finley Breese Morse, who received the first United States patent for a practical telegraph on June 20, 1840, No. 1,647.

The first long-distance communications were probably performed by runners and smoke signals. In the 1790s Claude Chappe, a Frenchman, introduced the semaphore as a means of long-distance communication. There was a semaphore line from Paris to Toulon having 120 semaphores placed six to ten miles apart. A single letter would take 10 to 15 minutes to make the trip. The first commercial semaphore system in the United States was installed by Jonathan Grout in 1800 on Martha's Vineyard to signal the approach of sailing ships.

The introduction of an electrical system of communication was initiated by the Italian Alessandro Volta in 1800, when he discovered that an electric current could be sent any distance through a wire. In 1819, Hans Christian Oersted, a Danish physicist, discovered the science of electromagnetism when he noticed that the needle of a compass wavered when he put it near a wire carrying an electric current. In 1825, William Sturgeon, an English electrician, demonstrated that an iron core strengthens a coil's magnetic field and built the first horseshoe-shaped electromagnet. In 1829, Joseph Henry, an American physicist, improved the Sturgeon electromagnet by producing a spool-wound magnet, thereby increasing the intensity. In 1833, Karl Friedrich Gauss, a German mathematician, in addition to making important contributions to the mathematical theory of electromagnetism, also discovered that an electric conductor could be reduced to a single circuit. Also in 1833, Wilhelm Weber, a German physicist, found that no insulation was necessary, except at the point of support, on wires carrying an electric current. In 1837, Sir Charles Wheatstone and Sir William Cooke patented a telegraph in England that worked by electromagnetism. It was used in England until 1870 when it was replaced by a much superior Morse telegraph system.

Samuel F. B. Morse was born in Charlestown, Massachusetts, on April 27, 1791. His father, Jedidiah Morse, was a Congregational minister and author of the best-known book on geography at that time. Young Morse was enrolled in Phillips Academy in Andover at the age of eight and Yale College at the age of 14. He graduated from Yale in 1810. While at Yale, Morse became interested in electricity under the influence of Professor Jeremiah Day and in

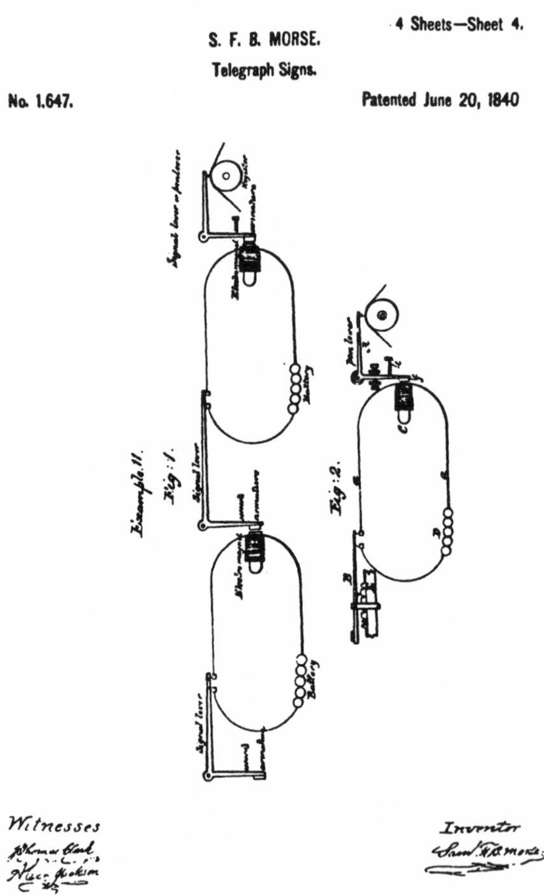

Samuel Finley Breese Morse's telegraph, patented June 20, 1840

painting as a pupil of Washington Allston.

After graduation, Morse went with Allston to London and remained there for four years studying under Allston. He was admitted to the Royal Academy, where he met and was befriended by Benjamin West. He won the gold medal of the Society of Arts for his terracotta statuette, Hercules.

In 1815, he returned to the United States and opened a studio in Boston, but no one seemed interested in his paintings. He was forced to make a living doing portraits, a task for which he had little enthusiasm. After three nonproductive years in Boston, he went to Charleston, South Carolina, at the invitation of his uncle, and achieved some success as a portrait painter. After a short stay in Concord, New Hampshire, where he met and married Lucretia Walker, he moved to New York in 1823. There, Morse and other artists founded the New York Drawing Association that led in 1826 to the founding of the National Academy of the Arts of Design. Morse enjoyed considerable popularity as president of the National Academy. Perhaps his best-known portraits were the two of Lafayette painted in 1825.

After the deaths of his wife in 1825, his father in 1826, and his mother in 1828, Morse became very discouraged. Thinking that a trip abroad might help, he went to Europe in 1829 and spent the next three years studying art and painting.

He returned to New York in October 1832 aboard the packet ship *Sully.* It was on this voyage that he met a fellow passenger, Charles Thomas Jackson, who demonstrated to Morse an electromagnet Jackson had acquired in Europe. That demonstration and the conversation that followed changed the direction of Morse's life and eventually led to the invention of the telegraph. During the remainder of the voyage of six weeks, he recorded in his notebook the essential electrical features of the invention and a dot-and-dash alphabet for sending and receiving messages, later modified and called the Morse code.

During the next few years, Morse taught painting at the University of the City of New York and worked on his telegraph. With the help and advice of electromagnet inventor Joseph Henry, chemist Leonard Gale, and assistant Alfred Vail, the telegraph was completed in January 1838. These men played a large part in the development of the telegraph, working on such essentials as the electromagnet, the relay, and the telegraph key. A caveat was filed with the United States Patent Office in September 1837, followed with an application for a patent on April 7, 1838. The patent was issued as No. 1,647 on June 20, 1840.

No one seemed interested in the telegraph for years, even though successful demonstrations had been carried out in New York, Philadelphia, and Washington. Finally, in 1843, Congress voted by a narrow margin $30,000 for an experimental line from Washington to Baltimore. Morse became superintendent at an annual salary of $2,500 and Vail as assistant superintendent at $3 per day, plus expenses. They set to work constructing this line, but a year and $23,000 later discovered that underground wires would not work. Fortunately, Morse hired Ezra Cornell, who later founded Cornell University, to help him in the construction of the line. Cornell, with Vail and Henry, suggested that the wires be placed above ground on poles using the necks of glass bottles as insulators. Morse finally agreed with this procedure.

The work was completed on May 23, 1844. One day later, Morse, sitting at a transmitter in the Supreme Court room in the Capitol in Washington sent a message to Alfred Vail in the Mount Clare station of the Baltimore and Ohio Railroad Company in Baltimore. The biblical message, "What hath God wrought?," suggested by Annie Ellsworth, daughter of the commissioner of patents, was received by Vail, who transmitted the same message back to Morse. Thus, the telegraph became a historical fact.

Morse offered to sell his invention to the government for $100,000, but his offer was refused. He then organized the Magnetic Telegraph Company and secured funds for the construction of a line from New York to Philadelphia. He hired Amos Kendall to manage his business and legal affairs. Little by little, telegraph lines began to multiply. By 1848, Florida was the only state east of the Mississippi that did not have the telegraph. Many companies sprang up. These infringed on Morse's patents, and he was forced to file many lawsuits to uphold his rights, which the courts always upheld. In 1854, the Supreme Court decided in favor of Morse as having been the original inventor of the telegraph. In 1856, Hiram Sibley and Ezra Cornell merged the systems they controlled and organized the Western Union Telegraph Company. On October 24, 1861, they put a line through to the west coast from St. Joseph, Missouri to Sacramento, California, causing the demise of the Pony Express. The company succeeded and the profits were enormous. Morse, through his patents, became a wealthy man.

Many honors were bestowed on Morse. Yale University conferred the LLD on him. European countries that had refused to grant him patents on his invention now gave him medals, money, and decorations. Vassar College came into existence with Morse as one of the founders. He was the first to introduce daguerrotypes in America, having learned the process from Louis Daguerre. In 1871, the telegraph operators of America placed a bronze statue of him in Central Park, New York. In 1900, Morse was selected as a charter member of the Hall of Fame for Great Americans on the campus of New York University.

In 1847, he built Locust Grove as a summer residence on a 200-acre estate on the Hudson River two miles below Poughkeepsie, New York. In 1859, he bought a brownstone house at 5 West 22nd Street in New York City as a winter residence. It was there that Morse died April 2, 1872; he was buried in Greenwood Cemetery in New York.

Telephone

The telephone patent is considered to be one of the most valuable patents ever issued. It was awarded to Alexander Graham Bell as United States Patent No. 174,465 on March 7, 1876.

The first telephone was the "string telephone" described by Robert Hooke, an English scientist, in 1667. It was made with a pair of cylinders, each of which had one end covered by a tight membrane. These were connected with a straight and taut string. Voices could be carried over a considerable distance by this method. In the 1840s, Hermann von Helmholtz, a German physicist, published a book entitled *Sensations of Tone*, in which were listed several experiments on the reproduction of sound with electrically driven tuning forks.

In 1849, Antonio Meucci of Florence, Italy, built an instrument capable of transmitting imperfect speech. This device was neither publicly demonstrated nor patented. On August 26, 1854, Charles Bourseul, of Paris, France, published an article in Volume XXIX of *L'Illustration* that described a method of transmitting sounds telegraphically by employing the principle of intermittent current. In 1861, Johann Philipp Reis, a school teacher from Friedrichsdorf, Germany, built the first telephone to be publicly demonstrated from a coil of wire, a knitting needle, the skin of a sausage, the bung of a beer barrel, and a strip of platinum. It was exhibited at the Physical Society of Frankfurt. This device also worked on the principle of intermittent current and was capable of producing only musical tones.

It was Alexander Graham Bell who solved the problem of transmitting articulate speech by varying the intensity of an unbroken current rather than working with an intermittent current, and for this he received the first United States patent for a telephone.

On the same day that Bell filed an application for his patent, Elisha Gray, a professor at Oberlin College, filed a caveat on the telephone with the Patent Office. Gray, as an American citizen, could file a caveat with the United States Patent Office whereas Bell, a British subject at that time, did not have that right and had to file an application on a completed invention. In view of this and the fact that the Patent Office record showed the fifth entry on February 14, 1876, was to Bell and the 39th entry on that day was to Gray, the Patent Office issued the first patent for a telephone to Bell.

Alexander Graham Bell was born on March 3,

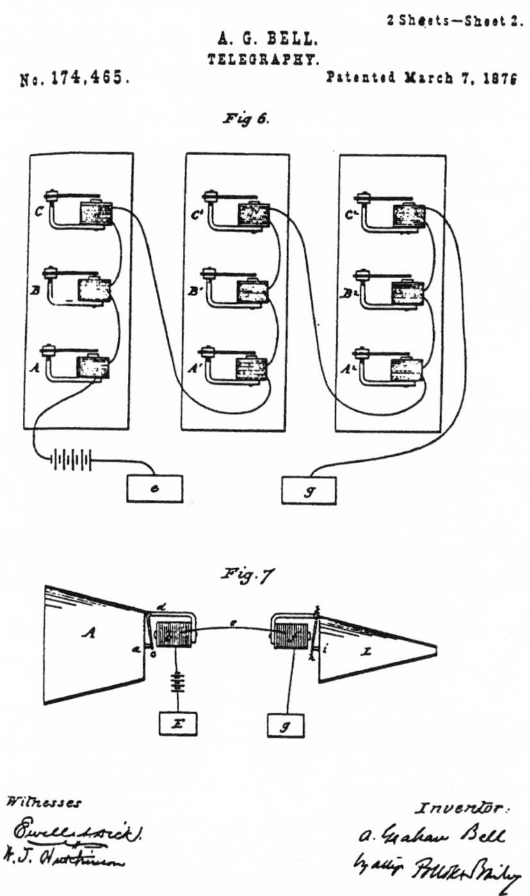

The telephone patent, issued to Alexander Graham Bell on March 7, 1876

1847, in Edinburgh, Scotland. His father, Alexander Melville Bell, was a pioneer teacher of speech to the deaf and the developer of a system called Visible Speech, in which written symbols represented certain basic sounds.

After being taught by his mother during his earlier years, young Bell was sent to McLauren's Academy in Edinburgh when he was 10 years of age. He graduated from the Royal High School at the age of 13. The next year he was allowed to visit his grandfather in London. His grandfather, the first Alexander Bell, was the founder of the art of elocution. The visit was extended to a year, and young Bell furthered his education in his grandfather's library and by accompanying him to his speech-making classes. At the end of the year, he returned to Edinburgh to become a pupil-teacher of elocution and music at the Weston House Academy in Elgin. He remained in Elgin until 1866, when he became an instructor at Somersetshire College in Bath, England.

Tragedy then struck the Bell family. Alexander Graham's brothers died of tuberculosis, and on the advice of doctors, the family moved to Canada in August 1870. After living in Brantford, Ontario, for two years, Alexander moved to Boston, where he opened a private school to which institutions could send teachers for training in the use of Visible Speech. In 1873, Bell was made professor of vocal physiology at Boston University.

In addition to teaching, Bell began experimenting on a "harmonic telegraph," an instrument that would have simultaneously transmitted multiple messages over a single wire. He sought technical assistance at the electrical shop of Charles Williams Jr. in Boston, where he was assisted by a young machinist named Thomas A. Watson. Bell and Watson spent many hours in Williams's workshop trying to perfect the harmonic telegraph. In an experiment on June 2, 1875, one of the transmitter springs of their telegraph stuck and as it was plunked generated a current that sent a faint noise that Bell was able to detect as not only the tone of the spring but also its overtones. This was the principle that he had been seeking. He knew that if an apparatus could produce overtones of a steel spring, then the human voice could be reproduced as well.

Two men who had a great influence on Bell and the telephone were Gardiner Greene Hubbard, a Boston attorney who later became Bell's father-in-law, and Thomas Sanders, a leather and hides merchant of Boston. These men agreed to finance the development of the telephone and formed the Bell Patent Association in which Hubbard, Sanders, Watson, and Bell would share in any benefits that might come as a result of these experiments.

During the fall of 1875, Bell and Watson continued to work on the telephone and on February 14, 1876, Bell filed his patent application at the Patent Office in Washington a few hours earlier than Elisha Gray's caveat. The patent was issued on March 7, 1876. The first intelligible sentence transmitted by telephone occurred on March 10, 1876, when Bell shouted, "Mr. Watson, come here, I need you!," after spilling a cup of acid on his clothes.

After further improvements, the telephone was exhibited at the International Centennial Exposition in Philadelphia in 1876. It was overlooked by the public until Don Pedro, Emperor of Brazil, became enthusiastic about the invention. The telephone then became the main feature of the Centennial.

In the fall of 1876, the Bell Patent Association offered to sell all rights to the telephone to the Western Union Telegraph Company for $100,000. The offer was rejected by William Orton, president of Western Union. On July 9, 1877, the Bell Patent Association became the Bell Telephone Company, a trusteeship with Hubbard as trustee and Sanders as treasurer. One of the major decisions that Hubbard made was that the company would lease telephones to licensed parties for a royalty. The company would sell service, not telephones. On February 17, 1879, the Bell Telephone Company joined with the New England Telephone Company to become the National Bell Telephone Company. In 1880, it became the American Bell Telephone Company. On December 30, 1899, the parent company of the Bell System became the American Telephone and Telegraph Company which had come into existence in 1885 as a subsidiary responsible for long-distance service. This company grew to be one of the largest in the world. Before divestiture on January 1, 1984, A T & T owned 182 million telephones, and more than a billion miles of wire, and employed nearly 1 million people.

Bell resigned from the company in 1880 after becoming both famous and rich. This enabled him to pursue both his first love, that of working for the deaf, and his second love, that of aviation. In addition, his interests also included the phonograph, a metal detector, an electric probe used in surgery, a prototype of the iron lung, an echo method of locating icebergs, a new metric system, and experiments in the scientific breeding of sheep that would result in the bearing of more than one lamb at a time.

Many honors came to Bell, including honorary degrees, gold medals, the presidency of the National Geographic Society, opening (with his friend and associate Watson) the first transcontinental telephone line from New York to San Francisco, and a regent of the Smithsonian Institution. But perhaps his greatest honor came on November 10, 1882, when he became a citizen of the United States. On his gravestone are the words: "Born in Edinburgh...died a citizen of the U.S.A."

Alexander Graham Bell died on August 2, 1922, at his estate on Cape Breton Island, Nova Scotia. He was buried on top of a mountain in a tomb cut in the rock in Beinn Bhreagh, near his Nova Scotia home.

Television

Millions of people watch television every day but few if any give thought to Vladimir Kosma Zworykin, who is considered "the father of television." On December 20, 1938, he received the first United States patent for an iconoscope, a cathode-ray transmitting tube. He also received a patent for a kinescope, a cathode-ray receiving tube. These components form the essential parts of modern television. The patent number for the iconoscope was 2,141,059.

Discoveries basic to television appeared long before Zworykin and developed along two lines: the mechanical system and the electronic system. Pioneers in mechanical television were Alexander Bain, John L. Baird, C. F. Jenkins, Herbert Ives, and Ernst Alexanderson. Pioneers in electronic television were C. R. Carey of Boston, who worked with a mosaic of selenium cells in 1875; Paul von Nipkov of Germany, who introduced the scanning disc in 1883; Ferdinand Braun of Germany, who developed the cathode-ray oscilloscope in 1897; Julius Elster and Hans Geitel of Germany, who improved the photoelectric cell in 1905; Boris Rosing of Russia, who proposed a system using a mechanical transmitter and a cathode-ray tube receiver in 1907; A. A. Campbell-Swinton of England, who proposed the use of cathode-ray tubes for both the transmission and reception of images in 1907; Howard W. Weinhart of the United States, who patented a miniature television tube in 1925; and Philo Farnsworth of the United States, who invented the image dissector tube in 1927. It was Zworykin's inventions of the iconscope and the kinescope that formed the heart of modern-day television.

Vladimir Kosma Zworykin was born in Mourom, Russia, on July 30, 1889. Following his graduation from high school, he received a degree in electrical engineering from the St. Petersburg Institute of Technology in 1912. In the institute at that time was a professor of physics, Boris Rosing, who introduced to Zworykin the application of electronics to the problems of television. Zworykin went to Paris for two years to study X-ray research under Professor Paul Langevin at the College de France. At the outbreak of World War I, he returned to Russia to serve in the Signal Corps of the Russian Army.

After the armistice, he traveled twice around the world, deciding that the United States would be his future home. He arrived in the United States in 1919

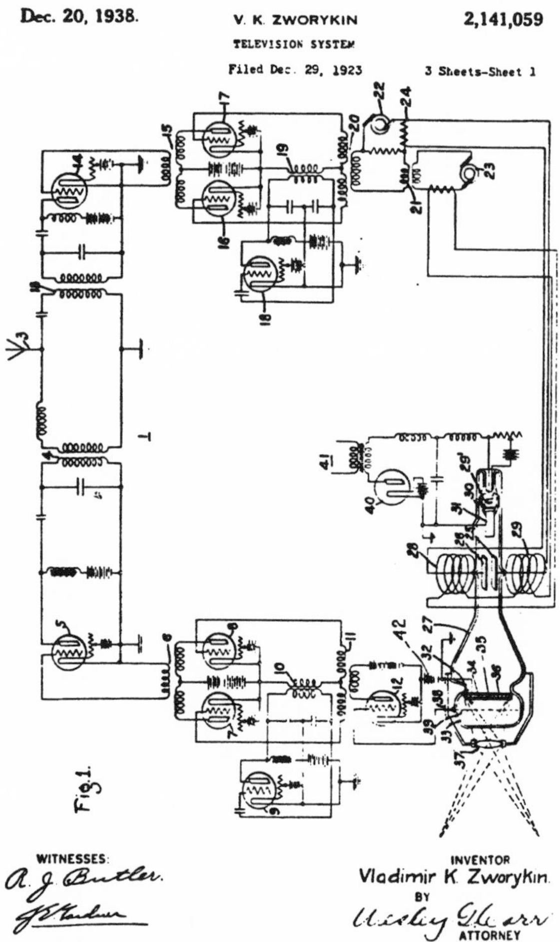

Zworykin's iconoscope patent made modern television possible

and found work as a bookkeeper in the Russian Embassy. In 1920, he accepted a position at the Westinghouse Electric Company in the research laboratory to work on the development of radio tubes and photoelectric cells. He left Westinghouse at the end of a year to establish a laboratory in Kansas for an oil company to investigate the cracking of gasoline by electrical methods. He returned to Westinghouse some eighteen months later. While at Westinghouse, he attended the University of Pittsburgh and received his PhD in 1926. His doctoral dissertation was *The Study of Photoelectric Cells and Their Improvement*. In 1924, he became a naturalized citizen of the United States.

On November 29, 1923, he filed a patent application on a cathode-ray transmitting tube, which he called the iconoscope. Interference proceedings with Philo Farnsworth prevented the patent from being issued until December 20, 1938. He also filed in 1924 a patent application on a cathode-ray receiving tube, which he called the kinescope. Zworykin first demonstrated his inventions to Westinghouse officials in 1924, and then publicly at a meeting of the Institute of Radio Engineers in Rochester, New York, on November 18, 1929.

In 1929, Zworykin moved from Westinghouse in Pittsburgh to the RCA Research Laboratory in Camden, New Jersey, to be director. He continued to work in the television field but had other interests as well. These interests included an infrared image tube, an electron microscope, the electric eye, a calculator, an oscillograph, an electronic diffraction camera, a clock without any moving parts, an electronic highway system, and an electronic reading device for the blind. He was awarded more than eighty United States patents.

Many awards came to Zworykin: the Overseas Award by the British Institute of Electrical Engineers in 1939; the Modern Pioneer Award by National Association of Manufactures in 1940; the Gold and Silver Rumford Medals in 1941; the honorary doctorate by the Brooklyn Polytechnic Institute in 1941; the War Department Certificate by the United States government in 1945; the Navy Certificate of Commendation by the United States government in 1947; the Scott Medal by the Franklin Institute in 1947; the vice president of RCA in 1947; Chevalier of the French Legion of Honor in 1948; the Presidential Certificate of Merit by the U.S. government in 1948; the Lamme Medal by the American Institute of Electrical Engineers in 1948; and the Poor Richard Club Gold Medal in 1949.

Vladimir Zworykin died July 29, 1982, in Princeton, New Jersey.

Transistor

The first United States patent for a transistor, No. 2,524,035, was awarded to John Bardeen and Walter H. Brattain on October 3, 1950. Although Bardeen and Brattain received the first patent, the story of the transistor would not be complete without including the work of their director, William Shockley.

Jons Berzelius, a Swedish chemist, discovered silicon in 1824, and Clemens Alexander Winkler, a German chemist, discovered germanium in 1886. These elements are from a group of materials called semiconductors, whose electrical properties are intermediate between those of metals and insulators. It was from a study of these materials by Shockley, Bardeen, and Brattain that the transistor was born in 1947. It was found that the device that they developed transmitted current across a resistor. The name transistor was first suggested by John Robinson Pierce, an American electrical engineer.

John Bardeen was born May 23, 1908, in Madison, Wisconsin. His father, Charles Russell Bardeen, was dean of the University of Wisconsin Medical School. Young Bardeen attended the public schools in Madison and graduated from the University of Wisconsin with a BS degree in 1928 and an MA degree in 1929, both in electrical engineering. In 1930, he was employed as a geophysicist at the Gulf Research Laboratories in Pittsburgh, Pennsylvania. In 1933, he left Gulf and entered Princeton University where he earned a PhD in mathematical physics in 1936.

Bardeen was appointed to an assistant professorship in physics at the University of Minnesota in 1938. During World War II, he was a physicist with the Naval Ordnance Laboratory in Washington, D.C. In 1945, he joined the Bell Telephone Laboratories at Murray Hill, New Jersey, as a research physicist and as a co-worker with Walter Brattain and William Shockley.

Walter H. Brattain was born on February 10, 1902, in Amoy, China, where his American parents were living at the time. His father, Ross R. Brattain, was a schoolteacher. The family moved to the state of Washington when Walter was very young. After graduating from the public schools in 1920, he enrolled at Whitman College in Walla Walla, Washington and graduated with a BS degree in 1924. He furthered his education by earning his MA degree from the University of Oregon in Eugene in 1926 and

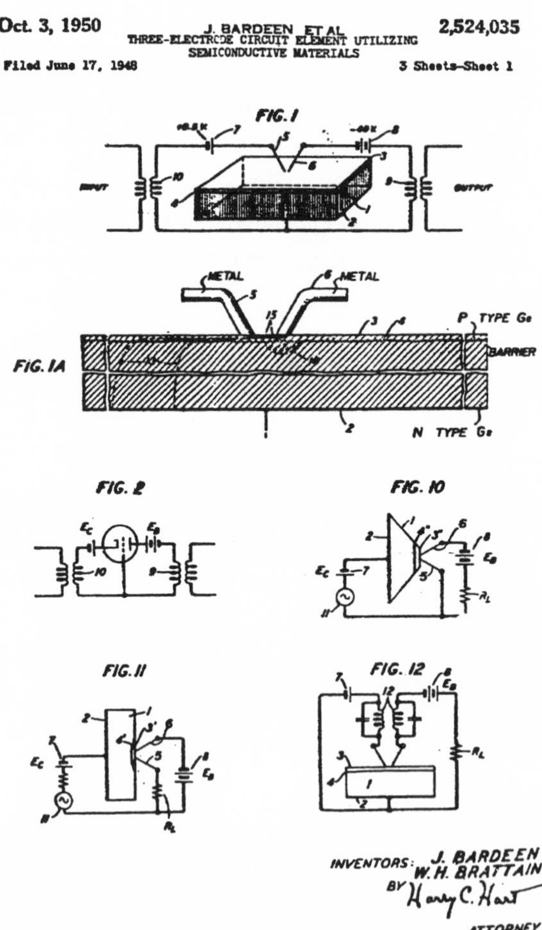

Bardeen and Brattain's transistor patent, issued October 3, 1950

his PhD degree from the University of Minnesota in Minneapolis in 1929.

For a short period, Brattain worked for the Bureau of Standards in Washington, D.C. and then joined the staff at the Bell Telephone Laboratories in 1929. He had been working in the field of thermionics for seven years when William Shockley joined the research staff at Bell. Both Brattain and Shockley began experiments on semiconductors, which work was interrupted in 1942 when both were assigned during World War II to do research at Columbia University in New York City on magnetic detection of submarines. When they returned to Bell Laboratories in 1945, they were joined by a new research member, John Bardeen.

William Bradford Shockley was born February 13, 1910, in London, England. His father, William Hillman Shockley, was an American mining engineer working in London when his son was born. When young Shockley was three years of age, the family moved to Palo Alto, California. He graduated from Hollywood High School in 1927 and enrolled in the California Institute of Technology, earning a BS degree in 1932. He received his PhD in physics from Massachusetts Institute of Technology in 1936 and immediately joined the Bell Telephone Laboratories, where Walter Brattain was working.

After serving with Brattain for the navy during the early part of World War II at Columbia University doing research on antisubmarine operations, Shockley became a consultant to the secretary of war from 1944 to 1945. He returned to Bell Laboratories to be joined by Bardeen and Brattain.

Shockley had previously in 1939 proposed using semiconductors as amplifiers to replace vacuum tubes. Starting with this proposal, Bardeen, Brattain, and Shockley began to work with semiconductors. The original experiments were conducted by Bardeen and Brattain and led to the development of the point-contact transistor in 1947 for which they received U.S. Patent No. 2,534,035. Shockley replaced the point-contacts with rectifying junctions in 1950, which resulted in the development of the junction transistor that eventually replaced the point-contact transistor. U.S. Patent No. 2,569,347 was awarded for this type of transistor. Shockley, Bardeen, and Brattain shared the 1956 Nobel prize for physics as a result of their transistor work.

The three Nobel scientists continued working with transistors at Bell Laboratories until 1951, when Bardeen was appointed as professor of electrical engineering and physics at the University of Illinois in Urbana, where he remained until he retired in 1975. He was the first person to receive two Nobel prizes in the same field, one for the transistor in 1956 and the other for low-temperature superconductivity in 1972. John Bardeen died of a heart attack January 30, 1991, in Boston, Massachusetts, while a patient at Brigham and Women's Hospital.

The two remaining Nobel physicists, Shockley and Brattain, continued working in solid-state physics at Bell Laboratories until 1955, when Shockley left to form the Shockley Semiconductor Laboratory in Palo Alto, California, the first semiconductor company in what was to become known as Silicon Valley. In 1963, he was appointed professor of engineering and applied science at Stanford University, where he remained until his retirement in 1975. His controversial views on human genetics were not well received by either the scientific community or the general public. Shockley died on August 12, 1989, at his home on the campus of Stanford University.

Walter Brattain remained at Bell Laboratories and did research on transistor physics until his retirement in 1967. He was then appointed as professor of physics at Whitman College. He died October 13, 1987, in Seattle, Washington.

Triode Vacuum Tube

One of the great inventions in the early twentieth century was the triode, an electron vacuum tube that magnifies weak signals without distortion. It was invented by Lee de Forest who received the first United States patent for an amplifier, No. 841,387, on January 15, 1907.

The age of electronics really began with Thomas A. Edison, who received the first patent in electronics, although at the time the field was not known by that name. In 1880, in an effort to eliminate the dark deposits on the inside of his electric lamp, Edison inserted a piece of wire into the bulb. He found that a current flowed from the hot filament to the wire probe. He was awarded U.S. Patent No. 307,031 on this novel lamp as a voltage indicator. This flow of current through a vacuum became known later as the Edison effect.

In 1897, Sir Joseph J. Thomson, a British physicist, discovered electrons and the fact that they carry a negative charge. In 1900, Owen W. Richardson, another British physicist, found that metals when heated emit electrons. Sir John A. Fleming, an English physicist and engineer, began experiments on the Edison effect. By 1905, he had made this phenomenon practical by discovering a device that would rectify high-frequency oscillations. He had invented the Fleming valve, or diode, which was able to convert alternating current to direct current, and for it he received U.S. Patent No. 803,684. But it was Lee de Forest in 1907, who turned the Fleming valve into an amplifier as well as a rectifier by inserting a grid of iron wires between the filament and the plate. This device, called the audion, or triode, was to be an important element in long-distance communications for many years.

Lee de Forest was born August 26, 1873, in Council Bluffs, Iowa. His father, Henry Swift de Forest, was the minister of the First Congregational Church in Council Bluffs. In 1879, the family moved to Talladega, Alabama, where the Reverend de Forest took a position as head of Talladega College.

Lee's father wanted him to study the classics in preparation for the ministry, but much to his father's dismay the boy was more interested in electricity than in books. After Lee convinced his father that he was better at science than Latin, his father agreed that he should attend the Sheffield Scientific School at Yale, his father's alma mater, after a year's prepara-

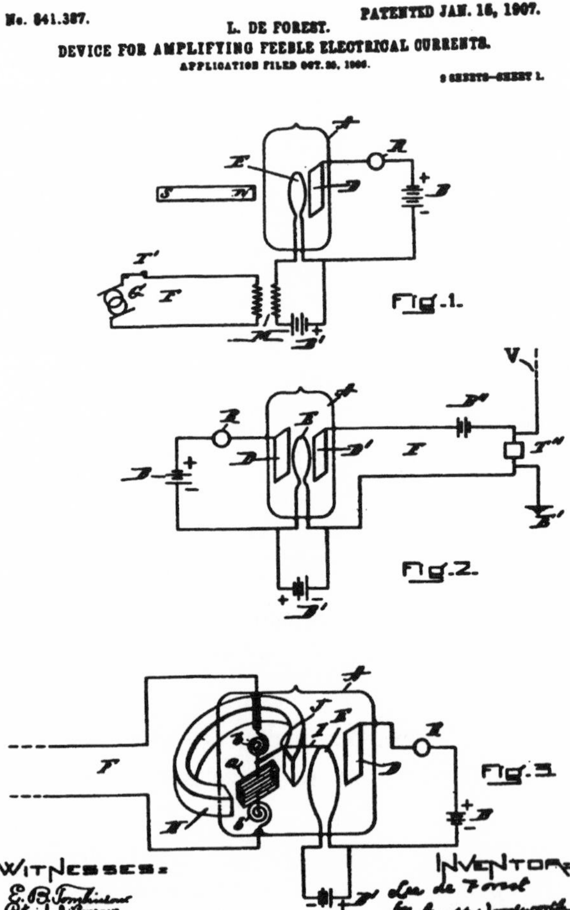

The triode, invented by Lee de Forest, and patented on January 15, 1907 magnifies weak signals without distortion

tion at the Dwight L. Moody Mount Hermon Boys School in Massachusetts.

Young de Forest entered Yale in 1893. After pursuing courses in physics and electricity, he graduated in 1896. He remained at Yale to do graduate work under J. Willard Gibbs. These studies were interrupted for a short time in 1898 when he joined the army during the Spanish-American War to become a bugler. After returning to Yale, he received his doctorate in 1899, his thesis centering on the recently discovered hertzian waves.

After graduation, de Forest took a job with the Western Electric Company in Chicago for $8 a week. He supplemented his income by editing the *Western Electrician*. His spare time was spent on developing a device for detecting wireless signals that he called the "Sponder." Long-distance tests with this detector were so successful that he and his partner, E. W. Smythe, decided to patent and market the device. Needing money and publicity, de Forest persuaded the *Publishers' Press* to give him a chance with his wireless system to report the 1901 International Yacht Races in New York in competition with Marconi and his system for the *Associated Press.* The broadcasts of both de Forest and Marconi failed, but it did bring the publicity that de Forest was seeking.

De Forest rented a machine shop on the Jersey City waterfront and erected a wireless station, the first of its kind in that area. In 1902, he met Abraham White, a stock speculator and promoter, who organized the American De Forest Wireless Telegraph Company, with White as president and de Forest as vice president. The company prospered with orders from the War Department, the Navy, and the Signal Corps until 1907 when de Forest discovered that the financial schemes of his backers were not to his liking. He resigned but was allowed to keep his patent rights to an invention that his assistant, Clifford Babcock, called the audion, later known as the triode.

While de Forest was associated with the Wireless Telegraph Company, he had been since 1904 secretly working on a device that would amplify wireless signals. He knew of the Edison effect and that John A. Fleming had taken Edison's light bulb with the extra electrode and turned it into a rectifier that allowed current to flow in only one direction. De Forest took the Fleming device and in 1906 inserted a zigzag piece of platinum wire, which he called a grid, between the filament and the metal plate. In doing so he discovered that the Fleming rectifier became an extremely efficient amplifier while keeping its rectifying properties.

In 1907, after receiving a patent on this invention, he formed the De Forest Radio Telephone Company. After months of further experimentation and successful public tests, the Navy ordered many of their ships to be equipped with the de Forest radiophone equipment, including the twenty naval vessels that cruised around the world in 1908-1909.

The company enjoyed some prosperity until 1912 when de Forest and the officers of his company were arrested for mail fraud as a result of some over optimistic advertising claims. The jury acquitted de Forest, though it convicted and sent to jail two of his directors. This legal problem, together with many infringement suits with his competitors, led to a financial shortage. In an effort to alleviate part of this problem, de Forest took a job as research engineer for the Federal Telegraph Company in San Francisco for $300 a month. While there, he continued to experiment with various circuits in order to increase the amplification of the audion. In 1913, de Forest sold the telephone amplifying rights of the audion to the American Telephone and Telegraph Company for $50,000. With this money he returned to New York City and revived his Radio Telephone Company in addition to setting up a laboratory at Highbridge. In 1917, the American Telephone and Telegraph Company paid de Forest more than $250,000 for the remaining rights, including radio, of the audion.

De Forest now turned his attention to the development of talking motion pictures. He was the first to produce sound on the same film that carried the picture and demonstrated his phonofilm, as he called it, in 1923 at the Rivoli Theatre in New York. He was too early in this field, and his phonofilm never reached commercial success.

In 1934, the Supreme Court upheld his claim to the invention of the regenerative circuit as a transmitter after years of litigation with Edwin H. Armstrong, inventor of the FM radio circuit. Also in 1934, de Forest founded the Lee de Forest Laboratories in Hollywood, California, where he worked out the principles of diathermy and did early research in television and radar.

De Forest received more than 300 patents but perhaps none as important as the triode. With this device the transcontinental telephone, radio, and television became practical. His triode remained the open door to the field of electronics until the invention of the transistor in 1947.

Lee de Forest died in Hollywood, California, on June 30, 1961.

Typewriter

The first United States patent for a typewriter that became a commercial success went to Christopher L. Sholes, Carlos Glidden, and Samuel Soule on June 23, 1868, No. 79,265.

Many attempts were made before 1868 to build a machine that would eliminate handwritten documents. The first patent for a typewriter was an English patent granted by Queen Anne to Henry Mill, a waterworks engineer, on January 7, 1714. The first practical typewriter was built by Pellegrine Turri, of Reggio Emilia, Italy, in 1808. There is nothing known about the machine, but letters typed on it are in the Reggio State Archives.

The first U.S. patent for a typewriter was issued to William Austin Burt, a surveyor from Detroit, Michigan, on July 23, 1829. The only model of this machine was destroyed in the 1836 fire of the Patent Office. A French patent was issued to Xavier Projean, of Marseilles, France, in 1833. The characters were on separate bars arranged in a circle, and each type struck a common center.

United States Patent No. 3,228 was awarded to Charles Thurber, of Worcester, Massachusetts, on August 26, 1843 for a typewriter with a carriage that held the paper and moved longitudinally to space the letters and rotated to change the lines. In 1849, Pierre Foucault, a blind teacher, received a French patent for a machine that embossed the letters so others who were blind could read them. Another typewriter for the blind was patented in the United States in 1856 by Alfred Ely Beach, editor of the *Scientific American*. John Pratt of Greenville, Alabama, was awarded a patent for a typewriter on August 11, 1868. This machine, called the Pterotype, was exhibited in London and reported in the *Scientific American* on July 6, 1867.

Christopher Latham Sholes was born February 14, 1819, on a farm near Mooresburg, Pennsylvania. His father, Orrin Sholes, had moved from Connecticut to the farm that he had received as a bounty for serving in the War of 1812. When young Christopher was four, the family moved to Danville, Pennsylvania. He attended Henderson's school in Danville until he was 14 and then for the next four years completed an apprenticeship on the *Danville Intelligencer*. The family moved to Green Bay, Wisconsin, where young Sholes worked as a journeyman printer for the *Wisconsin Democrat* before becoming the state printer.

In 1839, Sholes moved to Madison, Wisconsin

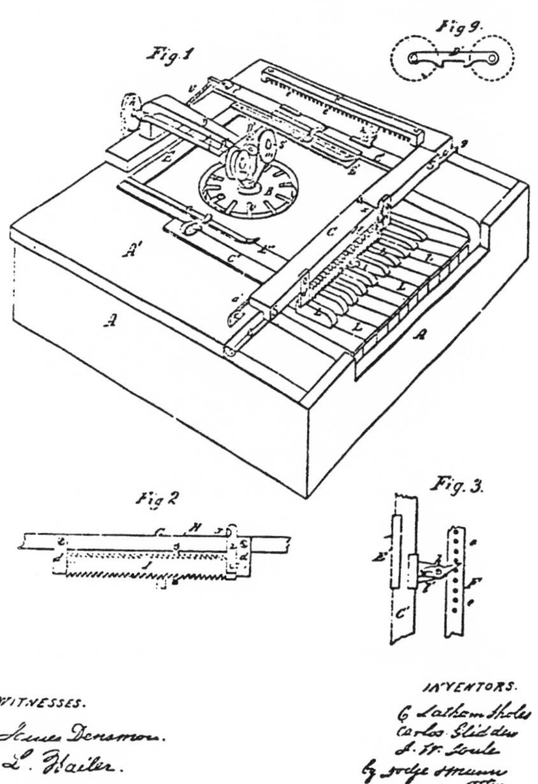

The patent for the first commercially successful typewriter, granted June 23, 1868

to become the editor of the *Wisconsin Enquirer*. The following year he became the editor of the *Southport Telegraph* in Southport, Wisconsin. He resigned four years later to become postmaster of Southport, an appointment made by President Polk. He later served two terms as state senator and one term in the state assembly. He moved to Milwaukee in 1860 to become editor of the *Milwaukee News* and later the *Milwaukee Sentinel*. During the Civil War, he was a loyal Unionist, and in 1863 President Lincoln appointed him as collector of the port of Milwaukee.

Sholes's inventive ability was evidenced while he was still in the newspaper business. He designed and built a machine that printed the names of suscribers on the margin of the newspapers. This machine was manufactured in 1860 in Madison, Wisconsin. Sholes's new job as port collector left him with time to pursue his inventive interests. He went often to the machine shop of Charles Kleinsteuber in Milwaukee to work on his inventions. It was there that he met Samuel Soule, a draftsman and civil engineer. Together, they built a paging machine patented on September 27, 1864, and a machine for numbering the pages of books, which was patented on November 13, 1866.

It was also in Kleinstuber's shop that Sholes met Carlos Glidden, who was working on a project of his own, a steam-driven rotary plow. Glidden pointed out to Sholes the article in the *Scientific American* for July 6, 1867, in which was described the typewriter exhibited in London by John Pratt. The article inspired Sholes to begin working on a letter-printing machine after enlisting the aid of Glidden and Soule, as well as Matthias Schwalbach, a machinist in the Kleinsteuber shop. The first machine produced in 1867 by Sholes and his associates was a crude affair, but it did work.

Sholes used his typewriter to send a letter to a friend, James Densmore, a wealthy businessman in Meadville, Pennsylvania. Densmore was so impressed with this typed letter that he became a partner, and from that moment he took command of the business. It was under his critical eye that many improvements were made to the machine. He also obtained the patent for Sholes, Glidden, and Soule on June 23, 1868. However, his manner and methods caused Glidden and Soule to drop out of the partnership eventually.

Between 1867 and 1873, Sholes and Densmore built fifty machines but none reached the stage of commercial acceptance. They then sought the assistance of G. W. N. Yost, a mechanical expert, who proposed various changes and suggested that the Remington Arms Company in Ilion, New York, could build the machines with the accuracy and skill needed. The Remingtons saw the possibilities in the typewriter and on March 1, 1873 proposed to Sholes and Densmore a settlement of $12,000 each or a royalty for their rights. They also insisted that the typewriter would bear the name Remington. Sholes accepted the cash, but Densmore took the royalty and later became rich. In 1882, Remington sold the typewriter rights to Wyckoff, Seamans & Benedict and by 1895 more than 200,000 had been made and sold.

Sholes continued working in Milwaukee on improving the typewriter, with the Remington company receiving all the results of his experiments. For the last nine years of his life, Sholes did not enjoy good health. Finally, on February 17, 1890, he died.

Vulcanized Rubber

Charles Goodyear received the first United States patent for vulcanized rubber. It was issued as patent No. 3,633 on June 15, 1844. The patent was reissued as Reissue No. 156 on December 25, 1849 and again on November 20, 1860 as Reissue Nos. 1,084 and 1,085.

Before Goodyear arrived on the scene, crude rubber had been used for centuries by the Indians of Central America. In 1730, Charles de la Condamine, a French explorer, introduced rubber to Europe. It became known as rubber after Joseph Priestley, a British chemist, noted in 1770 that the substance would erase pencil marks when rubbed over them.

The first rubber patent was granted to Jacob Frederick Hummel of Philadelphia, Pennsylvania, on April 29, 1813, for a method of waterproofing shoes with a varnish of elastic gum. In 1823, Charles MacIntosh, a Scottish manufacturer, made for the market waterproofed cloth by spreading a thin coating of rubber between layers of cloth. In 1832, F. Leudersdorff, a German chemist, first observed and reported the effect of sulfur on rubber.

Charles Goodyear was born in New Haven, Connecticut, on December 29, 1800. His father, Amasa Goodyear, was a hardware manufacturer, inventor, and merchant. Young Charles attended the public schools of New Haven and had the desire to study for the ministry. However, his father convinced him to enter the business world and helped him secure a job in a hardware store in Philadelphia to learn the merchandising trade. Charles remained in Philadelphia for four years and then returned to New Haven to become a partner in his father's business. Father and son worked together until 1826 when Charles opened a hardware store of his own in Philadelphia. Four years later he went bankrupt, owing thousands of dollars.

The first rubber company in America was established in 1832 in Roxbury, Massachusetts, and was called the Roxbury India Rubber Company. In 1834, Goodyear visited the Roxbury India Rubber Company, where he learned that the rubber companies were on the verge of collapse because the products made from rubber had serious faults. They softened and decomposed in summer and became hard and brittle in cold weather. The Roxbury manager told him that if someone would perfect a process of curing rubber to correct these faults, not only would that person become rich but would save an entire industry. This impressed Goodyear so much that he determined to be that person.

UNITED STATES PATENT OFFICE.

CHARLES GOODYEAR, OF NEW YORK, N. Y.

IMPROVEMENT IN INDIA-RUBBER FABRICS.

Specification forming part of Letters Patent No. 3,633, dated June 15, 1844.

To all whom it may concern:

Be it known that I, CHARLES GOODYEAR, of the city of New York, in the State of New York, have invented certain new and useful Improvements in the Manner of Preparing Fabrics of Caoutchouc or India-Rubber; and I do hereby declare that the following is a full and exact description thereof.

My principal improvement consists in the combining of sulphur and white lead with the india-rubber, and in the submitting of the compound thus formed to the action of heat at a regulated temperature, by which combination and exposure to heat it will be so far altered in its qualities as not to become softened by the action of the solar ray or of artificial heat at a temperature below that to which it was submitted in its preparation—say to a heat of 270° of Fahrenheit's scale—nor will it be injuriously affected by exposure to cold. It will also resist the action of the expressed oils, and that likewise of spirits of turpentine, or of the other essential oils at common temperatures, which oils are its usual solvents.

The articles which I combine with the india-rubber in forming my improved fabric are sulphur and white lead, which materials may be employed in varying proportions; but that which I have found to answer best, and to which it is desirable to approximate in forming the compound, is the following: I take twenty-five parts of india-rubber, five parts of sulphur, and seven parts of white lead. The india-rubber I usually dissolve in spirits of turpentine or other essential oil, and the white lead and sulphur also I grind in spirits of turpentine in the ordinary way of grinding paint. These three articles thus prepared may, when it is intended to form a sheet by itself, be evenly spread upon any smooth surface or upon glazed cloth, from which it may be readily separated; but I prefer to use for this purpose the cloth made according to the present specification, as the compound spread upon this article separates therefrom more cleanly than from any other.

Instead of dissolving the india-rubber in the manner above set forth, the sulphur and white lead, prepared by grinding as above directed, may be incorporated with the substance of the india-rubber by the aid of heated cylinders or calender-rollers, by which it may be brought into sheets of any required thickness; or it may be applied so as to adhere to the surface of cloth or of leather of various kinds. This mode of producing and of applying the sheet caoutchouc by means of rollers is well known to manufacturers. To destroy the odor of the sulphur in fabrics thus prepared, I wash the surface with a solution of potash, or with vinegar, or with a small portion of essential oil or other solvent of sulphur.

When the india-rubber is spread upon the firmer kinds of cloth or of leather it is subject to peel therefrom by a moderate degree of force, the gum letting go the fiber by which the two are held together. I have therefore devised another improvement in this manufacture by which this tendency is in a great measure corrected, and by which, also, the sheet-gum, when not attached to cloth or leather, is better adapted to a variety of purposes than when not prepared by this improved mode, which is as follows: After laying a coat of the gum, compounded as above set forth, on any suitable fabric I cover it with a bat of cotton-wool as it is delivered from the doffer of a carding-machine, and this bat I cover with another coat of the gum—a process which may be repeated two or three times, according to the required thickness of the goods. A very thin and strong fabric may be thus produced, which may be used in lieu of paper for the covering of boxes, books, or other articles.

When this compound of india-rubber, sulphur, and white lead, whether to be used alone in the state of sheets or applied to the surface of any other fabric, has been fully dried, either in a heated room or by exposure to the sun and air, the goods are to be subjected to the action of a high degree of temperature, which will admit of considerable variation—say from 212° to 350° of Fahrenheit's thermometer, but for the best effect approaching as nearly as may be to 270°. This heating may be effected by running the fabrics over a heated cylinder; but I prefer to expose them to an atmosphere of the proper temperature, which may be best done by the aid of an oven properly constructed with openings through which the sheet or web may be passed by means of suitable rollers. When this process is performed upon a

Charles Goodyear's patent for improvement in India-Rubber Fabrics, granted June 15, 1844

Goodyear returned to Philadelphia and began to experiment with rubber. For the next three years, using the kitchen of his home as a workshop, he mixed rubber with every material at his disposal. Some of his experiments were carried out in jail, where he had been sent for failing to pay his debts. The products he made with his mixtures were no better at the end of three years than those at the beginning.

In 1837, Goodyear moved his family to New Haven and then traveled on to New York to continue his experiments with the help of a friend who gave him a room and a druggist who supplied him with materials. It was not very long until he had some success by using nitric acid with copper or bismuth. For this process he obtained U.S.Patent No. 240 on June 17, 1837. He found a financial backer and began to make products with his patented process in an abandoned rubber factory on Staten Island, New York. However, these products never had much commercial success.

In 1838, Goodyear went to Roxbury, where he met Nathaniel M. Hayward, a former manager of one of the idle rubber companies. Hayward had discovered that sulfur dissolved in turpentine and mixed with rubber would eliminate its stickiness when exposed to sunlight. This was the first stage of vulcanization, and it remained for Charles Goodyear to perfect and patent the process. A short time later, Goodyear accidentally dropped a lump of the sulfur and rubber mixture on the top of a hot stove and observed that it charred but did not melt. Goodyear knew that he had at last solved the rubber problem. The answer was sulfur, rubber and heat.

For the next five years, Goodyear continued experimenting until he obtained the proper mix and heat for a uniform product. Finally, on June 15, 1844, he received a patent for the process and product of vulcanization. By now he was heavily in debt, having borrowed more than $50,000 for his experiments. However, he did license his process but at a price far below the fair market value. He never received enough money from his patent to repay this and other debts.

Poverty prevented Goodyear from filing for a British patent. He had tried to sell his process to Charles MacIntosh, a Scottish rubber manufacturer previously mentioned, but was turned down because he had no British patent. Samples left with MacIntosh were examined by Thomas Hancock, who detected the presence of sulfur. Upon further experimentation, Hancock independently discovered the vulcanization process and obtained a British patent on November 21, 1843, two months before Goodyear applied. A court judgment was rendered in favor of Hancock.

In 1851, Goodyear took his family to Europe where he staged a $30,000 rubber exhibit at the International Exhibition in London. All the items in the exhibit were made of rubber. In 1853, he wrote a book in two volumes entitled, *Gum Elastic and Its Varieties.* In 1855, he staged another rubber exhibit at the Exposition Universelle in Paris that cost him $50,000 of borrowed money. For this exhibit he received the Grand Medal of Honor and the Cross of the Legion of Honor. He heard about these honors while in debtor's prison at Clichy, France.

In 1858, Goodyear scraped together enough money for passage back to the United States for him and his family by pawning what few valuables the family possessed. After arriving back in his own country, he continued to experiment on new uses for his rubber, obtaining some 60 patents. Although he found nearly 500 uses for his rubber, he overlooked what was perhaps the most important one: rubber tires. Charles Goodyear died in New York City on July 1, 1860, leaving debts of nearly $200,000.

Xerography

The first United States patent, No. 2,297,691, for an electrophotographic copier was awarded to Chester F. Carlson on October 6, 1942.

The first copier was the human hand. Some early copying was done by means of inked wooden blocks carved by hand. In the 1450s, Johannes Gutenberg of Germany invented printing with movable type. This enabled any number of copies to be made of the same text. The invention of carbon paper in 1806 by Ralph Wedgwood of England and the typewriter in 1868 by Christopher Sholes, Carlos Glidden, and Samuel Soule of the United States produced copies at the same time as the master. Thomas A. Edison patented in 1876 a stencil duplicator called autographic printing. The hectograph, invented by Alexander Shapiro of Germany in 1880 and the mimeograph, invented by A. B. Dick of the United States in 1887, were used to make copies in the early part of the 20th century. Contact copying (the blueprint), projection copying (the photostat), heat-sensitive copying (Thermofax), the diazo system (Ozalid), and the dye transfer system (Verifax) were also used. The first photocopier on the market was the Rectigraph, patented by George C. Beidler of Oklahoma City, Oklahoma, in 1906. However, the greatest advance in the photocopying field was an invention in 1937 by Chester F. Carlson that depended on the use of photoelectric or photoconductive materials and plain dry paper.

Chester F. Carlson was born in Seattle, Washington, on February 8, 1906. His father, crippled by arthritis, worked as a barber when his condition allowed him to. Shortly after Chester was born, the family moved to Mexico, Arizona, and finally to San Bernadino, California, where young Chester grew up and attended elementary and high school. He then took a cooperative course at Riverside Junior College for three years before transferring to the California Institute of Technology, where he received his BS degree in physics in 1930.

Carlson took a job as a Research Engineer at the Bell Telephone Laboratories in New Jersey in 1930 and soon transferred to the patent department. He remained with Bell Labs until 1933, when he was dismissed, along with many others as the result of a national business downturn. He soon found work writing patent applications in a patent attorney's office near Wall Street. A year later, Carlson accepted a position in the patent department of the electronics firm called P. R. Mallory & Company in New York and attended the New York Law School

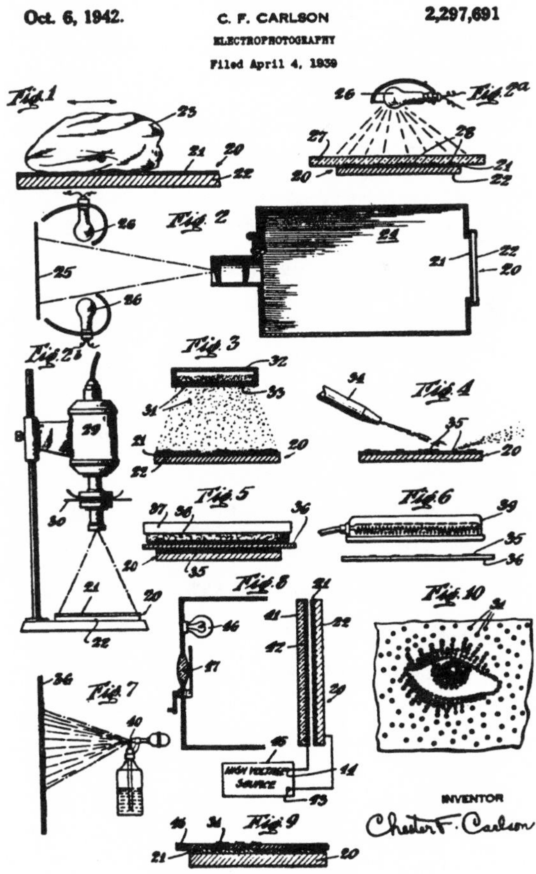

Chester Carlson advanced the field of photocopying with this patent, granted October 6, 1942

at night, receiving his law degree in 1939.

While working at Mallory, Carlson saw the need for some kind of fast copying machine for his patent applications because the conventional methods at that time were both time-consuming and expensive. In addition to his work, he began to research the subject at the New York Public Library. After months of reading, he came across a reference to the fact that the electrical conductivity of certain materials can be affected by exposure to light. Upon further reading, he discovered that sulfur was one of these materials. He started experimenting with this material in his apartment until the neighbors convinced him that this was not the proper place for this kind of laboratory.

Carlson hired Otto Kornei, an unemployed German physicist, as his assistant in 1938 and rented laboratory space above a bar in Astoria, Queens. It was not long until Carlson and his assistant produced the world's first successful xerographic copy on October 22, 1938, using a zinc plate coated with sulfur. The notation "10-22-38 Astoria" was written in black ink on a glass slide. The sulfur-coated zinc plate was rubbed with a handkerchief to give it an electrostatic charge. The glass slide was placed on the plate and exposed to a bright light for a few seconds. After the slide was removed, lycopodium powder was applied to the surface of the plate and the excess removed by gently blowing, leaving a duplicate of the notation in powder on the plate. The powder image was transferred to wax paper by pressing the paper against the plate. The image was made permanent by melting the wax with heat.

Otto Kornei remained with Carlson for six months after this first copy before accepting a job offer from IBM. Carlson continued to work on his copier and tried to find a company interested in his invention. From 1939 to 1944, more than 20 companies rejected his idea. Finally, in 1944, Carlson signed an agreement with the Battelle Memorial Institute, a private research foundation in Columbus, Ohio, in which Battelle agreed to develop his idea in exchange for 75 percent of any future royalties. The project was placed under the direction of Roland M. Schaffert, a Battelle research physicist, who during the next two years improved the copier to the extent that industry began to take notice.

One of the companies that became interested was the Haloid Corporation of Rochester, New York, a firm specializing in photographic materials. In 1947, a demonstration persuaded Joseph C. Wilson, Haloid's president, and John Dessauer, Haloid's chief of research, to sponsor further research at Battelle in exchange for the right to manufacture any products as a result of this research. Two years later, Haloid introduced its first copier called Model A. This machine proved to be too slow and complicated and not very dependable. However, in 1950, Haloid came out with the 914 Copier that eventually became a great success.

In 1948, Haloid and Battelle began to search for a name that better described the electrophotographic process. They adopted the name xerography, suggested by a professor at Ohio State University. In 1958, Haloid became Haloid-Xerox, and in 1961, the Xerox Corporation.

Carlson continued as a consultant at Haloid until royalties from the 914 made him a very wealthy man. He retired and enjoyed life in Rochester, New York, until his death on September 19, 1968.

Zipper

Whitcomb L. Judson received the first United States patent for a zipper on August 29, 1893. The number was 504,038.

The first fastening device was probably a straight pin made from bone. Later came lacings with eyelets or hooks, buckles, and buttons with buttonholes. In 1843, Charles Atwood was granted the first United States patent, No. 2,978, for hooks and eyes, which became popular in the 19th century. Although safety pins had been around for some time, Walter Hunt received the first United States patent, No. 6,281, for a safety pin in 1849. In 1883, Louis Hannart invented the snap. Many of these fasteners were placed in rows, and all had to be fastened and unfastened by hand, one after the other. It was Whitcomb L. Judson who revolutionized the fastening technique by inventing a simple mechanical means for quickly fastening and unfastening a series of connecting members, a device that was later called the zipper.

Little is known about the life of Whitcomb Judson except that he was a mechanical engineer and inventor in Chicago, Illinois. He had lived in Minneapolis and New York before moving to Chicago. From 1889 to 1892, he received 14 patents relating to street railways. He organized the Judson Pneumatic Street Railway Company of Minneapolis, Minnesota, and supervised the building of experimental railways in New York City and Washington, D.C. From 1889 to 1901, he was awarded additional patents for internal combustion engines, variable speed transmissions and clutches for automobiles, and a tread for traction wheels. Judson received all together 30 patents.

Colonel Lewis Walker, a corporation lawyer of Meadville, Pennsylvania, was associated with the Judson Pneumatic Street Railway Company. He was acquainted with Judson and in 1893 became interested in one of Judson's inventions, a new type of fastener. A year later, the Universal Fastener Company of Chicago was formed to sell the Judson fastener, with Colonel Walker as one of the stockholders. The company moved from Chicago to Elyria, Ohio, from Elyria to Catasauqua, Pennsylvania, and finally in 1900 to Hoboken, New Jersey. The first fasteners were handmade and sold under the name Universal. A machine for manufacturing these fasteners was built by the Manville Brothers of Waterbury, Connecticut, but proved to be so complicated that it was abandoned after a few years.

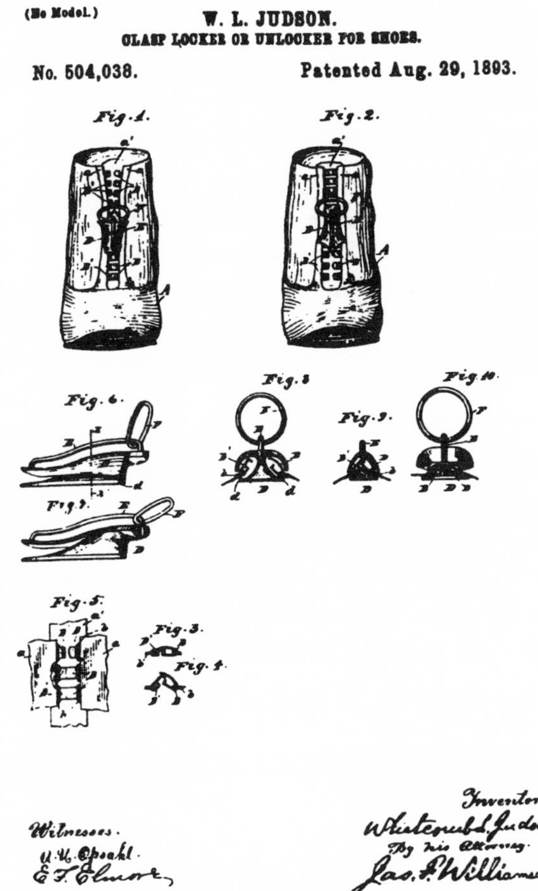

Whitcomb L. Judson's "Clasp Locker or Unlocker for Shoes," known to us as the zipper, was patented on August 29, 1893

In 1905, Judson designed an improved fastener that could more easily be machine-made. The company was reorganized as the Automatic Hook and Eye Company of Hoboken, New Jersey. The new company began to make and sell these redesigned fasteners under the name C-curity. Most sales were made door-to-door because the garment industry refused to use these fasteners on their clothing.

In 1906, Gideon Sundback, a Swedish engineer, was employed by the Automatic Hook and Eye Company as draftsman and design engineer. He improved the fastener and its method of manufacture. The improved model, called the Plako, was placed on the market in 1908. The garment manufacturers still refused to accept the product, so this model was also sold door-to-door. This sales method could not provide the company with enough money to sustain the business. It would have gone bankrupt had not the machine shop and engineering facilities taken in other work.

Sundback continued to work on improving the fastener and in 1913 invented a new type of slide fastener and the machinery to produce it. In 1913, Colonel Walker formed the Hookless Fastener Company, moved the plant to Meadville, Pennsylvania, and began to produce this new fastener, known as Hookless No. 2, which in principle is the same as the zipper we know today.

Sales came slowly for the company during the next few years. It was not until 1917 that a tailor in Brooklyn ordered a large quantity of slide fasteners to be used on money belts. The following year a contractor ordered 10,000 fasteners to be placed on flying suits for the U.S. Navy. In 1923, the B. F. Goodrich Company ordered and placed 150,000 Hookless fasteners on their galoshes marketed under the trademark Zipper Boots. The Zipper Boots marked the turning point in the success of the Hookless Fastener Company. However, the garment industry still resisted until 1934, when this type of fastener became an accepted part of men's trousers.

In 1928, the Hookless Fastener Company changed the name of its fastener to Talon, and, in 1937, changed the name of the company to Talon, Inc. It was acquired by Textron, Inc. in 1968.

Whitcomb Judson did not live long enough to see his invention become a worldwide indispensable item. He died in 1909.

Bibliography

Abbot, Willis J. *Aircraft and Submarines*. New York: G. P. Putnam Sons, 1918.

Acheson, Edward G. *A Pathfinder*. Port Huron, MI: Acheson Industries, Inc., 1965.

Adams, Robert McC. "Smithsonian Horizons." *Smithsonian* 18 (July 1987): 12.

Ahnstrom, D. N. *The Complete Book of Helicopters*. Cleveland, OH: World Publishing Company, 1954.

"Air Brake: George Westinghouse Jr." *Journal of the Patent Office Society XXX* (February 1948): 111-120.

"Air Conditioning." *Encyclopedia Americana*. 1964. Vol. 1.

Allen, James T., comp. *Digest of Cycles or Velocipedes With Attachments Patented in the United States From 1789 To 1892*. Washington, D.C.: U.S. Government Printing Office, 1892.

Allyn, Robert Starr. *The First Plant Patents*. Brooklyn, NY: Educational Foundations, Inc., 1934.

"Aluminum." *World Book Encyclopedia*. 1960. Vol 1.

American Association of Nurserymen. *Plant Patents*. Washington, D.C.: American Association of Nurserymen, Inc., 1957.

American Brake Shoe and Foundry Company. *The Story of the Brake Shoe*. New York: American Brake Shoe and Foundry Company, 1938.

American Dictionary of Printing and Bookmaking. New York: Howard Lockwood, Publisher, 1894.

"Ammonia." *Encyclopedia Americana*. 1988. Vol. 1.

Anderson, Leland I. *Bibliography: Dr. Nikola Tesla*. 2nd ed. Minneapolis, MN: The Tesla Society, 1956.

Angus-Butterworth, Lionel M. *The Manufacture of Glass*. New York: Pitman Publishing Corporation, 1948.

"Artillery." *World Book Encyclopedia*. 1960. Vol. 1.

Asimov, Isaac. *Asimov's Biographical Encyclopedia of Science and Technology*. Garden City, NY: Doubleday & Company, 1964.

—. *Chronology of Science and Discovery*. New York: Harper & Row, 1989.

"Automobile." *World Book Encyclopedia*. 1960. Vol. 1.

Baekeland, Leo H. "The Career of a Research Chemist." *Journal Franklin Institute* 230 (August 1940): 159-161.

Baida, Peter. "Breaking the Connection." *American Heritage* 36 (June/July 1985): 65-80.

—. "Eli Whitney's Other Talent." *American Heritage* 38 (May/June 1987): 22-23.

"Beekeeping." *Encyclopaedia Britannica*. 1957. Vol. 3.

Berle, Alf K., and L. Sprague De Camp. *Inventions, Patents, and Their Management*. New York: D. Van Nostrand Company, 1959.

Bigelow-Hartford Carpet Company. *A Century of Carpet and Rug Making in America*. New York: Bigelow-Hartford Carpet Company, 1925.

Biographical Directory of the American Congress 1774-1949, House Document 607. Washington, D.C.: U.S. Government Printing Office, 1950.

Birch, Robert L. "Numbering of U.S. Patents: Abortive New Sequence Begun in 1861." *Journal of the Patent Office Society* XLVIII (July 1966): 425.

Birge, Raymond T. "Presentation of the Nobel Prize to Professor Ernest O. Lawrence." *Science* 91 (April 5, 1940): 324-330.

Bitting, Arvill W. *Appertizing or the Art of Canning*. San Francisco, CA: The Trade Pressroom, 1937.

Blakeslee, Howard W. "Atomic Slingshot." *Science Digest* (April 1949): 63-67.

Bolles, Albert S. *Industrial History of the United States*. Norwich, CT: The Henry Hill Publishing Company, 1878.

Boorstin, Daniel J. *The Americans: The Democratic Experience*. New York: Random House, 1973.

"Bottle." *World Book Encyclopedia*. 1969. Vol. 2.

Brown, Donald L. "Protection Through Patents: The Polaroid Story." *Journal of the Patent Office Society* XLII (July 1960): 439-455.

Bruns, Roger, and Bryan Kennedy. "The Dream of Yesterday Is the Reality of Tomorrow." *American History Illustrated* XXIV (Summer 1989): 24-29, 48.

Bryant, Lynwood. "The Silent Otto." *Technology and Culture* 7 (1966): 184-200.

Bugbee, Bruce W. *Genesis of American Patent and Copyright Law*. Washington, D.C.: Public Affairs Press, 1967.

Bullock, Alan, and R. B. Woodings, eds. *20th Century Culture: A Biographical Companion.* New York: Harper & Row, 1983.

Burlingame, Roger. *Engines of Democracy*. New York: Charles Scribner's Sons, 1940.

Business Office Training Course of Addressograph Company. Chicago: Addressograph-Multigraph Corporation, 1937.

Butler, Joseph G. Jr. *Fifty Years of Iron and Steel.* Cleveland, OH: The Penton Press, 1922.

Byers, J. Harold. "The Selden Case." *Journal of the Patent Office Society* XXII (October 1940): 719-736.

Byrn, Edward W. *The Progress of Invention in the Nineteenth Century*. New York: Munn & Company, 1900.

Campbell, Hannah. *Why Did They Name It?* New York: Ace Books, 1964.

Carpenter, Rolla C., and Herman Diederichs. *Internal Combustion Engines: Their Theory, Construction, and Operation*. 3rd ed. New York: D. Van Nostrand Company, 1900.

Carruth, Gorton. *The Encyclopedia of American Facts and Dates*. 2nd ed. New York: Thomas Y. Crowell Company, 1959.

Carter, E. F. *Dictionary of Inventions and Discoveries*. New York: Crane, Russack & Company, 1976.

Chinn, George M. *The Machine Gun*. Vol. 1. Washington, D.C.: U. S. Government Printing Office, 1951.

Clark, E. H. "The Past of the Patent Office." *Journal of the Patent Office Society* XIV (April 1932): 262-272.

Clark, Ronald W. *Works of Man*. New York: Viking Penguin, Inc., 1985.

Clark, Victor S. *History of Manufactures in the United States*. Vol. 1. New York: McGraw-Hill Book Company, 1929.

Clarke, Basil. *The History of Airships*. London: Herbert Jenkins, Ltd., 1961.

Clarke, D. K. *Exhibited Machinery of 1862*. London: Day & Son, 1864.

Cochrane, Charles H. *The Wonders of Modern Mechanism*. Philadelphia: J. B. Lippincott Company, 1896.

Cochrane, Robert. *The Romance of Industry and Invention*. London: W & R Chambers, Ltd., 1896.

Collings, Gileart H. *Commercial Fertilizers*. New York: McGraw-Hill Book Company, 1955.

Collins, A. Frederick. *A Bird's Eye View of Invention.* New York: Thomas Y. Crowell Company, 1926.

Collins, James H. *The Story of Canned Food*. New York: E. P. Dutton & Company, 1924.

Condit, Carl W. "Sullivan's Skyscrapers." *Technology and Culture* 1 (1959-1960): 84-87.

Cook, Robert C. "The First Plant Patent." *Journal of the Patent Office Society* XIV (May 1932): 398-399.

Cooke, David C., and Martin Caidin. *Jets, Rockets and Guided Missiles*. New York: The McBride Company, 1951.

Cooke, Donald E. *Marvels of American Industry*. Maplewood, NJ: Hammond, Inc., 1962.

Cooper, Grace Rogers. *The Invention of the Sewing Machine*. Washington, D.C.: Smithsonian Institution, 1968.

"Copying Machines." *Encyclopedia Americana*. 1988. Vol. 7.

Cortada, James W. *Historical Dictionary of Data Processing Biographies*. New York: Greenwood Press, 1987.

Coulson, Thomas. "Some Prominent Members of the Franklin Institute. 6. Elihu Thomson,1853-1937." *Journal Franklin Institute* 264 (August 1957):87-103.

Crowther, J. G. *Discoveries and Inventions of the 20th Century*. London: Routledge and Kegan Paul, Ltd., 1966.

Current Biography. New York: The H. W. Wilson Company. See "Bardeen, John" (1957); "Birdseye, Clarence" (1946); "Brattain, Walter" (1957); "de Forest, Lee" (1941); "Fermi, Enrico" (1945); "Land, Edwin" (1981); "Sikorsky, Igor" (1946); "Townes, Charles" (1963); "Wright, Orville" (1946); and "Zworkin, Vladmir" (1949).

Current, Richard N. *The Typewriter and the Men Who Made It*. Urbana, IL: University of Illinois Press, 1954.

Dahn, Frank W. "Colonial Patents in the United States of America." *Journal of the Patent Office Society* III (March 1920): 342-349.

Darrow, Floyd L. *Masters of Science and Invention.* New York: Harcourt, Brace & Company, 1923.

De Bono, Edward. *Eureka!, An Illustrated History of Inventions From the Wheel to the Computer.* New York: Holt, Rinehart and Winston, 1974.

De Camp, L. Sprague. *The Heroic Age of American Invention.* Garden City, NY: Doubleday & Company, 1941.

Diamond, Freda. *The Story of Glass.* New York: Harcourt, Brace & Company, 1953.

Dictionary of American Biography. Vols. I-XX and Supplements One to Seven. New York: Charles Scribner's Sons, 1928-1981.

Dinsdale, Alfred. "Chester F. Carlson: Inventor of Xerography." *Photographic Science and Engineering* 7 (January-February 1963): 1-4.

Doolittle, William H. *Inventions of the Century. The Nineteenth Century Series.* Vol. XVI. Philadelphia: The Bradley-Garretson Company, 1903.

Doyle, Jack. "The First Plant Patent." *Rodale's Organic Gardening* 32 (November 1985): 24-32 and (December 1985): 62-68.

Drepperd, Carl W. *American Clocks and Clockmakers.* Garden City, NY: Doubleday & Company, 1947.

Duncan, Phillip S. *Amateur Radio* CQ 44 (January 1988): 50-52.

Dunlap, Orrin, Jr. *Radio's 100 Men of Science.* New York: Harper & Brothers, 1944.

Dutton, William S. *One Thousand Years of Explosives.* Philadelphia: The John C. Winston Company, 1960.

Eberle, Irmengarde. *Famous Inventors.* New York: Dodd, Mead & Company, 1941.

Eissler, Manuel. *The Modern High Explosives: Nitro-Glycerine and Dynamite.* New York: John Wiley & Sons, 1914.

"Electric Motor." *Encyclopedia Americana.* 1984. Vol. 19.

Emme, Eugene, M. *A History of Space Flight.* New York: Holt, Rinehart & Winston, 1965.

—. *The History of Rocket Technology.* Detroit, MI: Wayne State University Press, 1964.

Faber, Edward. *Nobel Prize Winners in Chemistry 1901-1961.* New York: Abelard-Schuman, 1963.

Fanning, Leonard M. *Fathers of Industries.* New York: J. B. Lippincott Company, 1962.

Fantel, Hans. "Against All Odds." *Opera News* 46 (August 1981): 16, 28.

Federico, Pasquale. "Anesthesia." *Journal of the Patent Office Society* VII (November 1924) 128-137 and (December 1924): 167-172.

—. "Colonial Monopolies and Patents." *Journal of the Patent Office Society* XI (November 1928): 598-605.

—. "Records of Eli Whitney's Cotton Gin Patent." *Technology and Culture 1* (1969-1970): 168-176.

—. "The Invention and Introduction of the Zipper." *Journal of the Patent Office Society* XXVIII (December 1946): 855-876.

—. "The Patent Office in 1837." *Journal of the Patent Office Society* XIX (December 1937): 954-957.

Feldman, Anthony, and Peter Ford. *Scientists & Inventors.* New York: Facts on File, Inc., 1979.

Finch, Volney C. *Jet Propulsion Turbojets.* Millbrae, CA: The National Press, 1948.

Fisher, Douglas Alan. *The Epic of Steel.* New York: Harper & Row, 1963.

Fuller, Edmund. *Tinkers and Genius.* New York: Hastings House, Publishers, 1955.

Garraty, John A., ed. *Dictionary of American Biography.* New York: Charles Scribner's Sons. See "Birdseye, Clarence." (Supplement Six, 1980).

Gartmann, Heinz. *Rings Around the World.* New York: William Morrow and Company, 1959.

Giedion, Siegfried. *Mechanization Takes Command.* New York: Oxford University Press, 1948.

Goodale, Stephen L., and J. Ramsey Speer. *Chronology of Iron and Steel.* Cleveland, OH: Penton Publishing Company, 1931.

Gould v. Mossinghoff. *United States Patent Quarterly* 229 (December 19, 1985): 1.

Graham-White, Claude, and Harry Harper. *The Aeroplane, Past, Present and Future.* Philadelphia: J. B. Lippincott Company, 1911.

Greenleaf, William. *Monopoly on Wheels.* Detroit, MI: Wayne State University Press, 1961.

Grierson, Ronald. *Electric Lift Equipment for Modern Buildings.* London: Chapman & Hall, Ltd., 1923.

Griffin, H. Hewitt. *Cycles and Cycling.* London: George Bell and Sons, 1903.

Grigson, Geoffrey, ed. *Things*. New York: Hawthorn Books, Inc., 1957.

"Gunpowder." *World Book Encyclopedia.* 1969. Vol. 8.

Hall, Cyril. *Triumphs of Invention.* London: Blackie and Son, Ltd., 1920.

—. *Seven Ages of Invention.* London: Blackie and Son, Ltd., 1931.

Hambleton, Ronald. *The Branding of America.* Dublin, NH: Yankee Books, 1987.

Hannon, Kerry. "Vindicated." *Forbes* (December 14, 1987): 35-36.

Hatfield, H. Stafford. *Inventions and Their Uses in Science Today*. London: Isaac Pitman & Sons, Ltd., 1939.

Hayes, Williams. *Chemical Pioneers*. New York: D. Van Nostrand Company, 1939.

Hays, Kathleen M. "Champion of the Mississippi." *Cobblestone* 11 (March 1990): 34-36

Hays, Will H. *See and Hear: A Brief History of Motion Pictures and the Development of Sound.* Reprinted. New York: Arno Press, 1970.

Hayward, L. H. "A Review of Helicopter Patents." *Supplement to Aircraft Engineering Journal* 5 (1952): 1-14.

Hecht, Jeff, and Dick Teresi. *Laser: Supertool of the 1980s*. New York: Ticknor & Fields, 1982.

Heyn, Ernest V. *A Century of Wonders: 100 Years of Popular Science*. Garden City, NY: Doubleday & Company, 1972.

Hill, Henry C., and Will H. Johnson. *The New Wonder Book of Knowledge*. Chicago: The John C. Winston Company, 1929.

Hilton, George W. *The Cable Car in America.* Berkeley, CA: Howell-North Books, 1971.

"History of the Cotton Gin." *The Pen and Lever* 1 (January 2, 1856: 4-5; (January 9 1856): 9-10; and (January 16, 1856): 1.

Hogan, Donald W. "Unwanted Treasures of the Patent Office." *American Heritage* 9 (February 1958): 16-19, 101-103.

Hogg, Garry. *Safe Bind, Safe Find: The Story of Locks, Bolts, and Bars*. London: Phoenix House, 1961.

Holbrook, Stewart H. *Machines of Plenty*. New York: The Macmillan Company, 1955.

Hopkins, Albert A. *Scientific American Reference Book.* New York: Munn & Company, 1905.

Hudson, Thomas B. "A Brief History of the Development of Design Patent Protection in the United States." *Journal of the Patent Office Society* XXX (May 1948): 380-399.

Hunziker, Otto Frederick. *Condensed Milk and Milk Powder*. 6th ed. La Grange, IL: Published by the Author, 1946.

Hyatt, John W. *Industrial and Engineering Chemistry* 6 (February 1914): 159.

Hylander, C. J. *American Inventors*. New York: The Macmillan Company, 1942.

Iles, George, C. J. *Leading Inventors*. New York: Henry Holt and Company, 1912.

Ingels, Margaret. *Willis Haviland Carrier: Father of Air Conditioning*. Garden City, NY: Country Life Press, 1952.

"Inventions of Dr. Elmer A. Sperry Based on the Gyroscope." *Journal Franklin Institute* 209 (May 1930): 669-677.

James, Edward T., ed. *Dictionary of American Biography*. New York: Charles Scribner's Sons. See "Goddard, Robert" and "Midgley, Thomas" (Supplement Three, 1974).

Jennings, W. N. "Frederic Eugene Ives: A Little Tribute To a Great Inventor." *Journal Franklin Institute* 225 (April 1938): 455-472.

Jerome, Chauncey. *History of the American Clock Business.* New Haven, CT: F. C. Dayton, Jr., 1860.

"Jet Engine." *Encyclopedia Americana.* 1988. Vol. 16.

"Jet Propulsion." *World Book Encyclopedia*. 1969. Vol. 11.

Jewkes, John, David Sawers, and Richard Stillerman. *The Sources of Invention*. New York: Macmillan & Company, 1958.

John Fritz Medal. New York: The John Fritz Medal Board of Award, 1917.

Johnson, Allen, ed. *Dictionary of American Biography.* New York: Charles Scribner's Sons. See "Bell, Alexander;" "Bigelow, Erastus;" "Blanchard, Thomas;" and "Borden, Gail" (Vol. II, 1929).

—. and Dumas Malone, eds. Dictionary of American Biography. New York: Charles Scribner's Sons. See "Davenport, Thomas" and "Eads, James" (Vol. V, 1930). See "Gatling, Richard"; "Glidden, Joseph"; and "Goodyear, Charles" (Vol. VII 1931). See "Haish, Jacob"; "Hall, Charles"; and "Hallidie, Andrew" (Vol. VIII, 1932).

Kaempffert, Waldemar. *A Popular History of American Invention*. Vols. I and II. New York: Blue Ribbon Books, 1924.

Kane, Joseph N. *Famous First Facts*. New York: The H. W. Wilson Company, 1964.

Keiper, Frank. *Pioneer Inventions and Pioneer Patents*. Rochester, NY: Pioneer Publishing Company, 1923.

Killeffer, D. H. *The Genius of Industrial Research*. New York: Reinhold Publishing Company, 1948.

Kirby, Richard S. *Inventors and Engineers of Old New Haven*. New Haven, CT: New Haven Colony Historical Society, 1939.

Klemm, Friedrich. *A History of Western Technology*. London: George Allen and Unwin, Ltd., 1959.

Knight, Edward H. *Knight's American Mechanical Dictionary*. New York: J. B. Ford and Company, 1874.

Kodak Milestones. Rochester, NY: Eastman Kodak Company, 1964.

Kursh, Harry. *Inside the U.S. Patent Office*. New York: W. W. Norton & Company, 1959.

Lake, Simon. "Modern Submarines in War and Peace." *International Marine Engineering* (July 1915 to April 1916): 1-16.

Lambermont, Paul, and Anthony Pirie. *Helicopters and Autogyros of the World*. Rev. ed. London: Cassell & Company, Ltd., 1970.

Lancaster, O. E., ed. *High Speed Aerodynamics and Jet Propulsion—Jet Propulsion Engines*. Princeton, NJ: Princeton University Press, 1959.

Larsen, Egon. *A History of Invention*. London: Phoenix House, Ltd., 1961.

—. *Ideas and Invention*. London: Spring Books, 1960.

—. *The True Book About Invention*. London: Frederick Muller, Ltd., 1954.

Larson, Cedric. "The Patent Office Models 1836-1890." *Journal of the Patent Office Society* XXXIII (April 1951): 243-253.

"Lathe." *Encyclopedia Americana*. 1988. Vol. 16.

Lavine, Sigmund A. *Famous Industrialists*. New York: Dodd, Mead & Company, 1961.

Leggett, M.D., comp. *Subject-Matter Index of Patents For Inventions Issued By the United States Patent Office From 1790 To 1873*, Inclusive. Vol. 1. Washington, D.C.: U.S. Government Printing Office, 1874.

Leithauser, Joachim G. *Inventor's Progress*. Cleveland, OH: The World Publishing Company, 1959.

Lengyel, Bela A. *American Journal of Physics* 34 (1966): 903-913.

Lent, Constantin Paul. *Rocketry: Jets and Rockets*. New York: Pen-Ink Publishing Company, 1947.

"Leopold Godowsky." *Encyclopedia of Practical Photography*. 1978. Vol. 7.

Lessing, Lawrence. *Man of High Fidelity: Edwin Howard Armstrong*. New York: J. B. Lippincott Company, 1956.

Lewis, Billy C. "The Bells Still Toll for San Francisco's Hills." *American History Illustrated* XXII (April 1987): 36-43.

Libbey-Owens-Ford Glass Company. *Glass: The Miracle Worker*. Toledo, OH: Public Relations Department, Libbey-Owens-Ford Glass Company, 1965.

Lincoln Electric Company. *Procedure Handbook of Arc Welding Design and Practice*. 11th ed. Cleveland, OH: The Lincoln Electric Company, 1957.

Liptrot, R. N., and J. D. Woods. *Rotorcraft*. London: Butterworths Scientific Publishers, 1955.

Livesay, Harold C. *American Made: Men Who Shaped the American Economy*. Boston: Little, Brown and Company, 1979.

Lookwood, Howard. *American Dictionary of Printing and Bookmaking*. New York: Howard Lookwood, Publisher, 1894.

MacDonald, George W. *Historical Papers on Modern Explosives*. New York: Whittaker & Company, 1912.

Maddox, Robert. "Fastest Man of the Earth and Skies." *American History Illustrated* XIX (June 1984): 10-17.

Mahadeva, K. "Eli Whitney: Pioneer of Mass Production." *The Chartered Mechanical Engineer* 13 (November 1966): 495-499.

Malone, Dumas, ed. *Dictionary of American Biography*. New York: Charles Scribner's Sons. See "Hoe, Richard"; "Holland, John"; "Howe, Elias"; and "Hyatt, John" Vol. IX, 1932). See "Kelly, William" and "Langstroth, Lorenzo" (Vol. X, 1933). See "McCormick, Cyrus" (Vol. XI, 1933). See "Mapes, James"; "Maxim, Hiram"; and "Mergenthaler, Ottmar" (Vol. XII, 1933). See "Morse, Samuel" and "Morton, William" (Vol. XIII, 1934). See "Otis, Elisha"; "Owens, Michael"; and "Painter, William" (Vol. XIV, 1934). See "Rodman, Thomas" and "Selden, George" (Vol. XVI, 1935). See "Sholes, Christopher" and "Sperry, Elmer" (Vol. XVII, 1935). See "Terry, Eli" (Vol. XVIII, 1936). See "Wright, Wilbur" and "Yale, Linus" (Vol. XX, 1936).

Maloney, John A. *Great Inventors and Their Inventions*. Chicago: University of Knowledge, Inc., 1940.

Manchester, Richard B. *Mammoth Book of Fascinating Information*. New York: A & W Visual Library, 1980.

Manning, Harold G. *Inventive America*. Waterbury, CT: Published by the Author, 1940.

Manual of Patent Examining Procedure. 5th ed. Washington, D.C.: U.S. Government Printing Office, 1983.

"Mark Air Conditioning's 50th Birthday This Year." *Heating, Piping and Air Conditioning* (February 1952): 84-87.

Marshall, Richard, ed. *Great Events of the 20th Century*. Pleasantville, NY: The Reader's Digest Association, Inc., 1977.

Martin, Thomas C., and Joseph Wetzler. *The Electric Motor and Its Application*. New York: W. J. Johnson, Publisher, 1887.

Marzall, John A. *The Story of the American Patent System, 1790-1952*. 2nd ed. Washington, D.C.: U.S. Patent Office, 1952.

Massucci, Edoardo. *Automobiles & Model Cars*. London: Orbis Publishing, Ltd., 1972.

Maurer, Allan. *Lasers: Light Wave of the Future*. New York: Arco Publishing, Inc.,1982.

Maxa. Rudy. "Let There Be Light." *The Washington Post Magazine* (February 12, 1978): 10, 13.

McCarthy, James R. *A Matter of Time*. New York: Harper & Brothers, 1947.

McMullen, J. C. "A Review of Patents on Silicon Carbide Furnacing." *Journal of the Electrochemical Society* 104 (July 1957): 462-465.

McMurtrie, Douglas C. *The Golden Book: The Story of Fine Books and Bookmaking—Past and Present*. Chicago: Pascal Covici, Publisher, 1928.

Meyer, Jerome S. *World Book of Great Inventions*. Cleveland, OH: The World Publishing Company, 1950.

Mitchell, Charles Eliot. *Proceedings of the Congress: Patent Centennial Celebration*. Washington, D.C.: Superintendent of Documents, 1891.

Moskowitz, Milton, Robert Levering, and Michael Katz. *Everybody's Business*. New York: Doubleday, 1990.

Mount, Ellis, and Barbara A. List. *Milestones in Science and Technology*. New York: Oryx Press, 1987.

Mumford, John Kimberly. *The Story of Bakelite*. New York: Robert L. Stillson Company, 1924.

National Cyclopaedia of American Biography. Vol. XLVI. New York: James T. White Company, 1963.

Nelson, Eric V. "History of Beekeeping in the United States." *Agriculture Handbook* No. 335. Washington, D.C.: U.S. Government Printing Office, 1971.

Neville, Leslie E., and Nathaniel F. Silsbee. *Jet Propulsion Progress*. New York: McGraw-Hill Book Company, 1948.

New York State Agricultural Society. *Report on the Trial of Plows Held at Utica, New York*. Albany, NY: Printing House of Van Benthuysen & Sons, 1868.

Nickerson, Stanton P. "Tetraethyl Lead: A Product of American Research." *Journal of Chemical Education* (November 1954): 560-571.

Norwig, E. A. "The Patents of Thomas A. Edison." *Journal of the Patent Office Society* XXXVI (March 1954): 213-232 and (April 1954): 275-296.

Nurnberg, John J. *Crowns: The Complete Story*. Paterson, NJ: Lont & Overcamp Publishing Company, 1967.

Obituary. "Edwin Land." New York: *New York Times*, March 2, 1991.

—. "Igor Sikorsky." Washington, D.C.: *Washington Star and News*, October 27, 1972.

—. "John Bardeen." New York: *New York Times*, January 31, 1991.

—. "Valdemar Poulsen." New York: *New York Times*, August 7, 1942.

—. "William Schockley." Johnson City, TN: *Johnson City Press*, August 15, 1989.

Oesper, Ralph E. "Christian Friedrich Schonbein. Part I. Life and Character." *Journal of Chemical Education* 6 (March 1929): 432-440 and (April 1929): 677-685.

Oliver, John W. *History of American Technology*. New York: The Ronald Press Company, 1956.

O'Neill, John J. *Prodigal Genius: The Life of Nikola Tesla*. Hollywood, CA: Angriff Press, 1981.

"Orthopaedic Appliances Atlas." *Artificial Limbs*. Vol. 2. Ann Arbor, MI: J. W. Edwards, 1960.

Owen, David. "Copies in Seconds." *The Atlantic Monthly* (February 1986): 65-72.

Palmer, Brooks. *The Book of American Clocks*. New York: The Macmillan Company, 1950.

Panati, Charles. *Panati's Extraordinary Origins of Everyday Things*. New York: Harper & Row, 1987.

Parton, James. *Triumphs of Enterprise, Ingenuity, and Public Spirit*. New York: Virtue & Yorston, 1872.

Palex Corporation et al. v. Mossinghoff et al. *United States Patent Quarterly* 225 (March 7, 1985): 243.

Pendray, G. Edward. *The Coming Age of Rocket Power*. New York: Harper & Brothers, 1947.

"Photography: History." *World Book Encyclopedia*. 1969. Vol. 15.

Polaroid Corporation v. Eastman Kodak Company. *United States Patent Quarterly* 228 (February 3, 1986): 305.

Popovic, Vojin, Radoslav Horvot, and Nikola Nikolic. *Nikola Tesla: Lectures, Patents, and Articles*. Beograd, Yugoslavia: Nikola Tesla Museum, 1956.

Postotnik, Pauline. "Anesthesia: A Long Way From Biting Bullets." *FDA Consumer* (June 1984): 24-27.

Pratt, Charles E. *The American Bicycler*. Boston: Houghton, Osgood and Company, 1879.

Putti, V. "Historical Artificial Limbs." *American Journal of Surgery* Vol. VI. New Series (January 1929): 111-117 and (February 1929): 246-253.

Rae, John B. *Automobile: A Brief History*. Chicago: University of Chicago Press, 1965.

Raymond, William Chandler. *Curiosities of the U.S. Patent Office*. Syracuse, NY: Wm. C. Raymond, Publisher, 1888.

Riley, John J. *A History of the American Soft Drink*. Washington, D.C.: American Bottlers of Carbonated Beverages, 1958.

Ringwalt, J. Luther. *American Encyclopedia of Printing*. Philadelphia: J. B. Lippincott & Company, 1871.

Roberts, John T. "A Reappraisal of the American System of Patent Examination." *Journal of the Patent Office Society* XLVIII (March 1966): 156-205.

Roberts, Royston M. *Serendipity: Accidental Discoveries in Science*. New York: John Wiley & Sons, Inc., 1989.

Robertson, Patrick. *The Book of Firsts*. New York: Clarkson N. Potter, Inc. Publisher, 1974.

Roe, Joseph Wickham. *English and American Tool Builders*. New York: McGraw-Hill Book Company, 1926.

Rogers, Albert Henry. *Firearms of Yesterday*. New York: Vantage Press, Inc., 1953.

Rosenblum, Naomi. *A World History of Photography*. New York: Abbeville Press, 1984.

Rossman, Joseph. "Plant Patent No. 1." *Journal of the Patent Office Society* XIII (October 1931): 521-525.

Rowley, J. F. *A Catalog of the Famous Rowley Artificial Leg*. Chicago: J. F. Rowley Company, 1922.

Rush, C. W., W. C. Chambliss, and H. J. Gimpel. *The Complete Book of Submarines*. Cleveland, OH: The World Publishing Company, 1958.

Sample, N. W., and T. Coulson. "The Story of the Bicycle." *Journal Franklin Institute*. 235 (April 1943): 393-404.

Sank, Jon R. "Oberlin Smith: Centenary of Magnetic Recording." *Audio* (December 1988): 90-94.

Sawyer, Charles Winthrop. *Firearms in America: The Revolver 1800 to 1911*. San Leandro, CA: Charles Edward Chapel, 1939.

Schurr, Cathleen, "Two Hundred Years of Patents and Copyrights." *American History Illustrated* XXV (July/August 1990) 60-71.

Schuyler, Robert L., and Edward T. James, eds. *Dictionary of American Biography*. New York: Charles Scribner's Sons. See "Carothers, Wallace"; "Ives, Frederic"; and "Thomson, Elihu" (Supplement Two, 1958).

Schwartz, Jacob R. *The Acrylic Plastics in Dentistry*. Brooklyn, NY: Dental Items of Interest Publishing Company, 1950.

Scott, John. *Genius Rewarded or the Story of the Sewing Machine*. New York: Book and Job Printer, 1880.

Shearer, Lloyd. "Nobel: The Man & His Prizes." *Parade Magazine* (December 11, 1977): 22.

Shiers, George. "The First Electron Tube." *Scientific American* 220 (March 1969): 104-112.

Silliman, Benjamin. *Electro-Magnetism: History of Davenport's Invention*. New York: G. & C. Carvill & Company, 1837.

Singer, Charles, ed. *A History of Technology*. Vol. 5. New York: Oxford University Press, 1958.

Sipley, Louis W. *Photography's Great Inventors*. Philadelphia: American Museum of Photography, 1965.

Smith, F. G. Walton. "Ships That Flew." *Sea Frontiers* 28 (July-August 1982) 203-216.

Smith, J. Bucknell. *A Treatise Upon Cable or Rope Traction*. New York: John Wiley & Sons, 1887.

Smith, William E. *Continued Study Units in Cultural Life: I. Colonial Inventions*. Philadelphia: F.A. Davis Company, Publishers, 1940.

Sorensen, Susan. "Evolution of a Dream." *Bicycling* (December 1988): 68-74.

Spencer, Charles. *Bicycles and Tricycles, Past and Present*. London: Griffith & Farran, 1884.

Starr, Harris E., ed. *Dictionary of American Biography*. New York: Charles Scribner's Sons. See "Acheson, Edward"; "Berliner, Emile"; "Burroughs, William"; "Curtiss, Glenn"; and "Eastman, George" (Supplement One, 1944).

Steinman, David B., and Sara Ruth Watson. *Bridges and Their Builders*. New York: Dover Publications, Inc., 1957.

Stern, Albert. "Galleries of Gismos." *Americana* (August 1986): 26-30.

Stern, Philip Van Doren. "Doctor Gatling and His Gun." *American Heritage* 8 (October 1957): 48-51, 105-108.

Steward, John F. *The Reaper*. New York: Greenburg, Publisher, 1931.

Stokley, James. *Science Remakes Our World*. New York: Ives Washburn, Publisher, 1942.

Thompson, John S. *The Mechanism of the Linotype*. Brookings, SD: Lebawarts Printers, 1948.

Thompson, Slason. *A Short History of American Railways*. Chicago: Bureau of Railway News and Statistics, 1925.

Time. Special Independence Issue 105, No. 20 (1975): 5-13.

Turck, J. A. V. *Origin of Modern Calculating Machines*. Chicago: The Western Society of Engineers, 1921.

Tylman, Stanley D., and Floyd A. Peyton. *Acrylics and Other Synthetic Resins Used in Dentistry*. Philadelphia: J. B. Lippincott Company, 1946.

United States Air Force Training Command. *Helicopter: History and Aerodynamics*. Randolph, TX: Headquarters Air Training Command, United States Air Force, January 1961.

United States Patent Model Foundation. Pamphlet. Alexandria, VA: 510 King Street, Suite 420.

Urquhart, John W. *Sewing Machinery*. London: Crosby Lockwood and Company, 1881.

Usher, Abbott P. *A History of Mechanical Inventions*. Boston: Beacon Press, 1929.

Van Gelder, Arthur Pine, and Hugo Schlatter. *History of Explosives Industry in America*. New York: Columbia University Press, 1927.

Viall, Ethan. *Electric Welding*. New York: McGraw-Hill Book Company, 1921.

Vivian, E. Charles, and W. Lockwood Marsh. *A History of Aeronautics*. New York: Harcourt, Brace & Company, 1921.

Walker, W. M. *Notes on Screw Propulsion: Its Rise and Progress*. New York: D. Van Nostrand and Company, 1861.

Wallace, Emily Duane. "Elihu Thomson: In Retrospect." *Journal Franklin Institute* 235 (May 1943): 479-481.

Wallechinsky, David, and Wallace Irving. *The People's Almanac*. Garden City, NY: Doubleday & Company, 1975.

—. *The People's Almanac # 2*. New York: William Morrow & Company, 1978.

Walton, Perry. *The Story of Textiles*. Boston: Walton Advertising & Printing Company, 1925.

Ward, Baldwin, ed. *A Pictorial History of Science and Engineering*. New York: Year, Inc., 1958.

Wasson, Tyler, ed. *Nobel Prize Winners*. New York: The H. W. Wilson Company, 1987.

Wellerson, Thelma L. "Historical Developments of Upper Extremity Prosthetics." *Orthopedic & Prosthetic Appliance Journal* (September 1962): 73-75.

Wensberg, Peter C. *Land's Polaroid: A Company and the Man Who Invented It.* Boston: Houghton Mifflin Company, 1987.

Whittle, Sir Frank. *Jet: The Story of a Pioneer*. New York: Philosophical Library, Inc., 1953.

Williams, Archibald. *The Romance of Modern Invention.* London: C. Arthur Pearson, Ltd., 1903.

Wilson, Mitchell. *American Science and Invention: A Pictorial History*. New York: Bonanza Books, 1954.

Woods, Clinton. *Ideas That Became Big Business*. Baltimore, MD: Founders, Inc., 1959.

Worrel, Rodney K. "The Wright Brothers' Pioneer Patent." *American Bar Association Journal* 65 (October 1979): 1512-1518.

Wright, Milton. *Invention and Patents*. New York: McGraw-Hill Book Company, 1927.

Yates, Raymond F. *Guide to Successful Inventing*. New York: Wilfred Funk, Inc., 1949.

Yost, Edna. *Modern Americans in Science and Invention.* New York: J. B. Lippincott Company, 1941.

Index

D

About the Author

Travis Brown was born in Washington County, Tennessee, in 1926. He was educated in the public school system of Washington County and obtained his B.S. and M.A. degrees in chemistry from East Tennessee State University in Johnson City, Tennessee.

After teaching in elementary school and working in an industrial laboratory, he became a patent examiner with the United States Patent and Trademark Office in Arlington, Virginia. He retired from this position in 1986.

He now lives with his wife, Maxine, in Johnson City, Tennessee, where both do volunteer work in the community. They have a son, David, and daughter-in-law, Maureen, who live in Kensington, Maryland.

Mr. Brown wrote *Historic First Patents* after retiring. This is his first book.